QUANTUM DOT SENSORS

edited by
John F. Callan
Françisco M. Raymo

Technology and Commercial Applications

QUANTUM DOT SENSORS

Pan Stanford Publishing

Published by

Pan Stanford Publishing Pte. Ltd.
Penthouse Level, Suntec Tower 3
8 Temasek Boulevard
Singapore 038988

Email: editorial@panstanford.com
Web: www.panstanford.com

British Library Cataloguing-in-Publication Data
A catalogue record for this book is available from the British Library.

Quantum Dot Sensors: Technology and Commercial Applications

ISBN 978-981-4316-00-2 (Hardcover)
ISBN 978-981-4364-61-4 (eBook)

Printed in the USA

Contents

Preface

The past two decades have witnessed the emergence of semiconductor quantum dots (QDs) as versatile nanoparticles with a wide range of applications from energy-efficient lighting and displays to biomolecular sensors. This book focuses on the incorporation of QDs into optical- and electrochemical-based sensing systems for chemical and biologically relevant target analytes. This arena has traditionally been dominated by organic-dye-based sensors. However, the realisation of the impressive optical properties of QDs has seen them emerge as serious rivals to their all-organic counterparts. QDs are now commercially available from many different vendors with a wide range of surface chemistries suitable for derivatisation.

In this book, we chart the progress in the development of QDs over the past two decades with a particular focus on the use of CdS-, CdSe- and CdTe-based QDs in sensing systems. Chapter 1 details the range of synthetic methods used to prepare QDs, including solvothermal, aqueous-based, microwave-assisted and micellar approaches with the benefits and potential drawbacks of each approach outlined. The preparation of biocompatible QDs is discussed in Chapter 2, which details how native hydrophobic QDs can be transferred into aqueous solution using methods such as ligand exchange and amphiphilic polymer encapsulation. Again, the advantages and disadvantages of the different approaches for the preparation of biocompatible QDs are outlined.

Modifying the surface of QDs with organic ligands or biomolecules is central to the development of QD-based sensors. This surface derivatisation can have a dramatic effect on the electrochemical properties of the semiconductor nanoparticle and affect its optical performance. Chapter 3 discusses the electrochemical properties of core and core–shell QDs and how these properties can change depending on the nature and thickness of the shell and the type of ligands attached to its surface. This chapter also details how the electroactivity of QDs can be exploited in the development of electrochemical-based sensing systems.

Although examples of electrochemical-based QD sensing are emerging, optical-based sensing methods still dominate the literature. The modulation of the QD optical signal in the analyte-free and bound states can be controlled by electron transfer or energy transfer mechanisms. Chapter 4 discusses how electron transfer between the QD and attached ligands or between the QD and the analyte itself (if the analyte is redox active) can be utilised to develop QD-based sensors. Chapter 5 details how QDs can be utilised as energy donors and integrated with a wide range of energy acceptors in energy transfer (FRET)-based sensors. Sensors for a wide range of analytes are discussed, ranging from simple protons to complex biomolecules. Finally, Chapter 6 discusses how differences in the core–shell composition of QDs and physical/chemical factors such as the size and nature of attached ligands can influence their toxicity in prokaryotic cells.

The aim of this book is to give the reader a detailed overview of how QDs can be prepared and incorporated as the signalling unit in electrochemical/optical-based sensors. The editors are indebted to the chapter authors for their contributions, the Pan Stanford publishing group and the Universities of Ulster and Miami for their assistance and support. We hope this book will be a useful reference text and will help in promoting QDs as viable alternatives in the design of optical sensors.

John F. Callan
Françisco M. Raymo
Winter 2012

Chapter 1

Quantum Dot Synthesis Methods

Yurii K. Gun'ko* **and Stephen Byrne**
School of Chemistry and CRANN,
Trinity College Dublin, Dublin 2, Ireland
*igounko@tcd.ie

1.1 Introduction

The synthesis of nanomaterials of controlled size and shape plays a crucial role in nanotechnology. Ever since primary investigations into their properties began in the 1980s, semiconducting nanocrystals (referred as quantum dots or QDs) have demonstrated a rich variety of properties, characteristics and potential applications. Microscale and larger semiconducting crystals have electronic and optical properties that depend on their bulk band structures. When the size of these crystals is reduced to the nanoscale, they essentially become "molecule-like" and this gives rise to unique properties which result in the nanocrystals differing greatly from their bulk counterparts [1–3]. Most importantly in nanocrystals certain properties are no longer influenced by their band structure but by size and shape [4, 5]. For example, in the case of CdS, the band gap can be tuned between 4.5 (275 nm) and 2.5 eV (450 nm) as the size

Quantum Dot Sensors: Technology and Commercial Applications
Edited by John F. Callan and Françisco M. Raymo

ISBN 978-981-4316-00-2 (Hardcover), 978-981-4364-61-4 (eBook)
www.panstanford.com

is varied from the macro to the molecular scale. The energy above the band gap required to add excess charge also decreases by 0.5 eV [6] and the melting temperature increases from 400 to 1600°C [7]. This range in the fundamental properties of a material opens up further possibilities to trace the evolution of electronic and optical properties of the matter from the bulk material to the atomic cluster [3, 8, 9].

The intensity of investigations into the synthesis and behaviour of QD nanocrystal systems has been remarkable over the past few decades. Initial contributions on the synthesis of CdS and ZnS were made by Brus and co-workers during the 1980s [10–12]. Moving on, other groups also continued with the production of CdS nanocrystals and comprehensively deciphered and analysed their unique properties [11, 13, 14]. Nozik *et al.* [2] investigated size quantisation in CdS and PbS nanocrystals, while Henglein and co-workers injected H_2S into a solution of cadmium perchlorate and sodium hexametaphosphate to give CdS nanocrystals with narrow size distributions [13]. The properties could be varied with pH and this led to the formation of a layer of $Cd(OH)_2$, which acted as an additional passivant for the removal of surface defect states. This shell was also attributed as the cause of dramatic luminescence increases without altering size when excess cadmium ions were added to a basic solution of CdS.

Here, we are going to consider most frequently used approaches to the synthesis of various QDs both in organic solvents and in aqueous medium.

1.2 High-Temperature Synthesis in Organic Solvents

1.2.1 Organometallic Synthesis

Most of the synthetic methods employed have been focused on obtaining highly crystalline, monodisperse and highly luminescent nanocrystals with desirable surface functionalities. Originally Bawendi *et al.* [15] were among the first to prepare CdE (E = S, Se, Te) nanocrystals, with a major emphasis on CdSe, *via* the use of dimethyl cadmium ($(CH_3)_2Cd$) and tri-*n*-octylphosphine selenide (TOPSe), which was then injected into a solution of the coordinating solvent tri-*n*-octylphosphine oxide (TOPO) at the high temperature (over 180°C). Slow growth and annealing in the coordinating solvent resulted in uniform surface derivatisation and regularity in the core structure, yielding CdSe nanocrystals with broad size ranges (1.2–11.5 nm). The ligands were exchanged using pyridine, precipitated with the addition of hexane, centrifuged and finally dispersed in a variety of organic solvents. This procedure enables the variability within the ensembles to be reduced down to <5%. Technically this method can be considered as a chemical co-precipitation, which involves a temporally discrete nucleation followed by slower controlled growth on the existing nuclei (Fig. 1.1). The colloidal system can also undergo so called Ostwald ripening, when the high surface energy of the small particles promotes their dissolution, then re-deposition of the material onto the larger particles enabling their further growth. Thus the nucleation and growth processes dictate the final particle size and morphology.

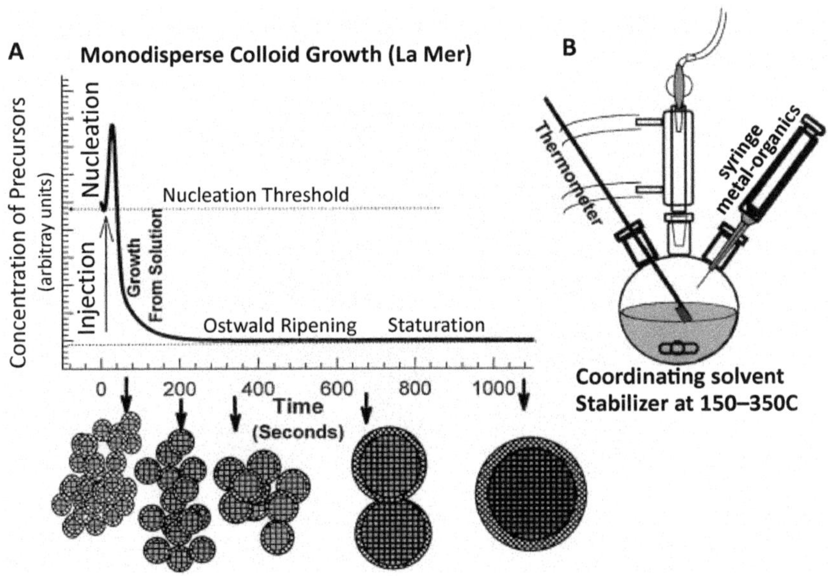

Figure 1.1 (A) The stages of nucleation and growth for the preparation of monodisperse NCs in the framework of the La Mer model. As NCs grow with time, a size series of NCs may be isolated by periodically removing aliquots from the reaction vessel. (B) Representation of the simple synthetic apparatus employed in the preparation of monodisperse NC samples. Reproduced from 16.

Further modification of this approach was developed by Alivisatos *et al.* [17]. The principal difference between the preparations was the injection temperature, which affects the time evolution of the nanocrystal growth as well as their crystallinity and shape. In this synthesis the TOPO was heated to 350°C under argon before injection (it was noted that this temperature was well above the flash point for TOPO and all manipulations were undertaken with extreme care) and after injection, the temperature was reduced to yield monodisperse ensembles. This change in temperature was utilised to distinctly separate both the nucleation and growth phases and resulted in high-quality crystalline, nearly monodisperse QD nanocrystals.

This work has been followed by other groups [18] and has been tweaked somewhat by Talapin *et al.* [19] in an effort to further reduce the amount of "defocusing" or Ostwald ripening occurring during the growth period (Fig. 1.2). This group prepared CdSe using a three-component hexadecylamine–trioctylphosphine oxide–triocty-lphosphine (HDA–TOPO–TOP) mixture which provided much better control over the growth dynamics.

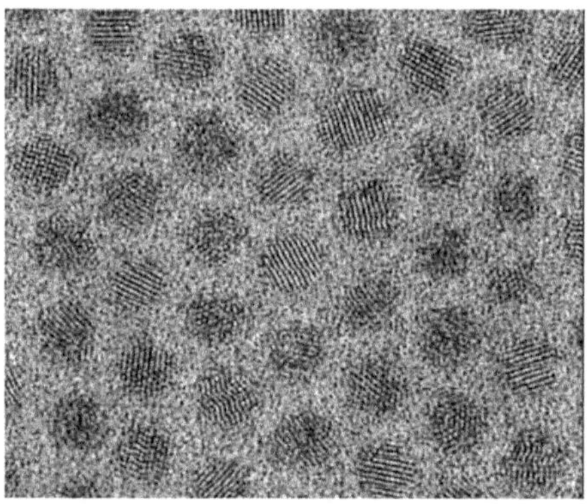

Figure 1.2 HRTEM image of monodisperse, highly crystalline CdSe nanocrystals, without any post-preparative treatment. Reproduced from 19.

This procedure, unlike some previous reports [15, 17, 20], allowed for the production of monodisperse CdSe nanocrystals without the need for any post-preparative treatment to narrow the broad size distributions. In contrast to the growth of CdSe nanocrystals in TOPO-TOP [21–23], a very fast "focusing" of the size distribution was observed with the HDA–TOPO–TOP system which continued throughout the entire growth process, while no "defocusing" (Ostwald ripening) was witnessed. This phenomenon was attributed to the large difference between the rate of nanocrystal growth in the "focusing" regime, which occurs due to the initial excess of the monomer (cadmium species) in solution, and the rate of the Ostwald ripening process.

An easier and "greener" approach was also been developed by Weller *et al.* [18]. They undertook a comparative study of the synthesis of CdSe QDs using the traditional but highly volatile $Cd(Me)_2$ and $Cd(Ac)_2$, a cheaper and more stable cadmium source and found differences in their reactivity. The researchers found that the $Cd(Me)_2$ to promote QDs growth *via* Ostwald ripening, while $Cd(Ac)_2$ promoted a more focussed growth. This separation of nucleation and growth has led to various in-depth investigations into the mechanisms involved during the synthesis of nanocrystals [20, 24, 25]. Peng *et al.* [20] investigated the factors that influenced the growth of II–VI (CdSe) and III–V (InAs) nanocrystals. They began their investigations based on two important facts, (1) diffusion-controlled growth leads to a narrowing of the size distributions with time [26] and (2) a variation in surface energy with size, provided the crystallites are small enough that Gibbs–Thompson effects come into play, will directly influence the growth kinetics [20, 25]. They define the two distinct regimes within the growth cycle. Initially the size distribution is "focussed" during the first few minutes and then "defocused" as the nanocrystals grow more slowly. They explain that the size distribution can be influenced by the Gibbs–Thompson effect on Ostwald ripening. At fixed monomer concentrations, the growth rate can be obtained by applying the Gibbs–Thompson equation:

$$S_r = S_b \exp\left(2\sigma V_m / rRT\right)$$

S_r/S_b = solubility of the nanocrystal and corresponding bulk

σ = specific surface energy

r = radius of the nanocrystal

V_m = molar volume of the material

R = gas constant

T = temperature

If $2\sigma V_m/rRT << 1$, the diffusion controlled growth of a nanocrystal of size r is

$$dr/dt = K(1/r + 1/\delta)(1/r^* - 1/r)$$

K = a constant proportional to the diffusion constant of the monomer

σ = thickness of diffusion layer

r^* = critical radius for which the solubility of the nanocrystals is equal to the monomer concentration in solution (zero growth rate)

According to this model, at any given monomer concentration there exists a critical size (r^*) that is in equilibrium with the precursor concentration in solution. During focussing, the nanocrystals present are all larger than r^* and the monomer concentration drops while particle concentration remains constant. Under these conditions smaller nanocrystals will grow at faster rates than the large ones.

During defocussing (Ostwald ripening), when the monomer concentration is depleted due to growth, r^* becomes larger than the average size present. Thus, the distribution broadens while smaller nanocrystals disappear or dissolute and larger ones continue to grow. During Ostwald ripening, different nanocrystals grow at different rates within the ensemble leading to a distinct broadening of the size distribution which manifests itself during characterisation as a broadening of the spectral (UV-vis absorption and fluorescence/photoluminescence (PL) emission) properties.

Precursor concentration also plays a crucial role. At higher concentrations, more nuclei are initially formed and these quickly expend the remaining monomers (cadmium containing species) as they grow. Thus, any remaining cadmium monomers induce an ineffective surface passivation of nanocrystal defects leading to a reduction in luminescence intensity. However, at low precursor concentrations, there is a reduction in the initial number of nuclei formed and cadmium monomers supplied to the surface effectively take part in ordering and reconstruction, leading to significant luminescence increases [27].

Talapin *et al.* [25, 28] followed on from this work and applied a theoretical model to explain the evolution of sizes within a nanocrystal ensemble. Their model took the size dependence of the activation energies of the growth and dissolution processes into account and they applied Monte Carlo simulations to establish conditions leading to either focussing or defocussing. The same group undertook additional experiments to determine the influence that the distribution of growth rates within the nanocrystal ensemble would have on the luminescence properties. They attributed a loss in quantum yield (QY) to a variation in the nanocrystal surface within the ensemble. Furthermore, they proposed that after a size-selective precipitation, the nanocrystal fraction with the highest quality or smoothest, defect-free surface corresponded to the crystal size with the smallest net growth rate during synthesis, (i.e., r^*, or zero growth rate as referred to previously) (Fig. 1.3).

This maximum emission intensity has further been investigated by Qu *et al.* [22]. They found that during the growth of CdSe nanocrystals, the QY increases to a certain maximum value and then decreases gradually. They denote the maximum as a luminescence "bright point" and attribute this point to being the optimum surface structure during nanocrystal growth. This emission maximum is highly dependant on the initial Cd:Se ratio and a large Se excess was required to achieve the most substantial results with narrowest size distribution.

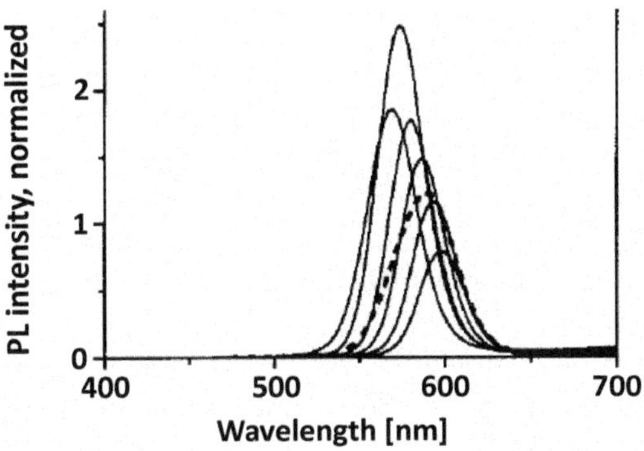

Figure 1.3 Luminescence emission spectra showing the QY evolution for each fraction of CdSe nanocrystals precipitated from a crude parent solution. Reproduced from 28.

1.2.2 QD Synthesis from Non-Organometallic Precursors

Overall, the production of QD nanocrystal *via* an organometallic route has led to the synthesis of high-quality semiconductor nanocrystals. However, some key chemicals used in this traditional route are extremely toxic, pyrophoric and explosive and can be very expensive. As considered above, CdSe nanocrystals are synthesised in a mixture of TOPO and TOP; however, the problem of irreproducibility of the growth dynamics and the shape of the

CdSe nanocrystals was mediated by the uncertain composition of the coordinating solvent. Technical grade TOPO (90%, Aldrich), for instance, provides better conditions for the growth of CdSe nanocrystals than distilled TOPO [19]. The solvent itself must be heated to extreme temperatures (300°C), which is well above its flash point giving rise to the risk of explosion or fire. Typically, $(CH_3)_2Cd$ is used and this is a highly dangerous and toxic compound. It is extremely air- and moisture sensitive and must be handled with great care and respect. In an effort to overcome this problem, variations such as CdO [29, 30], $Cd(Ac)_2$ [18] and cadmium myristate [31] have been used.

A number of alternative routes using safer and less expensive cadmium precursors and ligands have been reported [22, 30, 32] and have further promoted the synthesis of CdSe nanocrystals to an easily adoptable level. Reiss *et al.* [33] and Li *et al.* [32] have employed CdO for the synthesis of high-quality CdSe/ZnSe core/shell nanocrystals without the use of the traditional pyrophoric organometallic precursors. Qu *et al.* [22] have explored further and used CdO for the production of not only CdSe but also CdS and CdTe. This group [34] have also demonstrated that contrary to the belief that the synthesis of high-quality CdSe nanocrystals was a delicate affair; a variety of different systems can be used. They used a one-pot approach and different solvents (fatty acids, amines, phosphine oxides and phosphonic acids) to produce high-quality CdSe nanocrystals.

There were several reports on the utilization of single-source precursors for modified high-temperature synthesis of II–VI nanocrystals. The main advantages of the single-source precursor approach include (i) the possibility to get both components, e.g. Cd and Se, in one air and moisture stable precursor; (ii) the opportunity to avoid volatile toxic and/or pyrophoric precursors; (iii) reduced chances of the incorporation of impurities into the nanoparticles (one precursor is easier to purify) and (iv) the possibility of low temperature processing and decomposition [35]. However, this technique requires extra synthetic steps to prepare and purify single-source precursor compounds. Also, the nanoparticles prepared from single-source precursors normally have poorer luminescence properties (e.g. lower QYs) and sometimes give polydisperse particles.

One of the more successful and reproducible series of single-source precursors were the bis(dialkyldithio-/diseleno-carbamato)cadmium(II) or zinc(II) compounds, $M(E_2CNR_2)$ (where M = Zn or Cd, E = S or Se, and R = alkyl). These precursors allowed for the preparation of relatively high-quality, defect free, mono-dispersed nanoparticles using injection techniques at moderate temperatures (e.g. around 200°C) [35, 36].

The other useful precursors include molecular clusters such as $(X)_4[M_{10}Se_4(SPh)_{16}]$ (where X is Li^+ or $(CH_3)_3NH^+$, M is Cd or Zn, and SPh is phenyl thiolate). These moieties, which are stable under ambient conditions, have been used as single-source precursors for the controlled preparation of 2–9 nm CdSe and CdSe/ZnS nanocrystals and 2–5 nm ZnSe nanocrystals at temperatures between 220 and 240°C, thus removing the pyrolytic step for nucleus formation. Consequently, better synthetic control, slower thermodynamic growth and high crystallinity while maintaining narrow size dispersity (around 5%) [37] were achieved.

It is important to mention that significant work has also been done on the synthesis of II–VI nanocrystals with a variety of elongated shapes (e.g. CdSe nanorods) [24] and nanostructures with the tetrapod-like morphology (Fig. 1.4) by organic phase reactions. These amazing nanostructures have been of great interest due to their potential use in photovoltaics [38], in the synthesis of asymmetric nano-assemblies [39] and prospective additives for polymer reinforcement [40]. The synthesis of nanorod and tetrapod-like quantum structures and the shape control of II–VI semiconductor nanomaterials has been thoroughly reviewed by Kumar and Nann [41]; therefore, we are not going to consider these materials in this chapter.

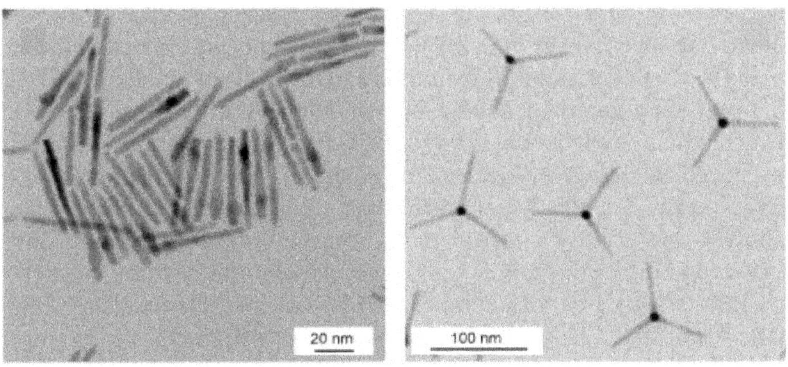

Figure 1.4 Selected TEM images of CdSe/CdS nanorods and nanotetrapods. Reproduced from 42.

1.2.3 Microwave Synthesis

Over recent years, the use of microwaves (MW) for the successful synthesis of high-quality QDs has increased and deserves particular attention. Microwave chemistry offers unique opportunities to significantly increase the rate of nanoparticle formation, while not compromising on the high-quality and narrow-size distributions needed for these nanomaterials to be further utilised in other areas. The microwave approach also provides the potential for a scale-up of the QD synthesis without the potential hindrance due to thermal gradient effects; something that is particularly important for further industrial large-scale production of nanoparticles. The utilisation of these techniques allowed for the preparation of high-quality, nearly monodispersed InGaP, InP, CdS, CdSe, CdTe and other QDs from appropriate molecular precursors in non-microwave absorbing alkane solvents [43–45]. Importantly, most chalcogenide precursors can selectively absorb the microwave energy, which enables the instantaneous nucleation and growth of the QDs upon MW irradiation. Therefore, using this method, QDs can be grown rapidly (no longer requiring elongated synthesis over several hours or even days) and controlled by a combination of reactant concentration and MW power, while QD size can be templated by the reaction temperature [44]. The microwave approach also allows the user to avoid the high-temperature injection procedure that significantly facilitates the synthesis of the previously mentioned QDs. For example, good quality CdSe nanocrystals have been prepared using the single-source precursor $Li_4[Cd_{10}Se_4(SPh)_{16}]$ by its microwave treatment (CEM Discover microwave) in aminohexadecane at a temperature range between 160 and 300°C. The same paper also reports the synthesis of monodisperse InGaP and InP QDs of high quantum efficiency (15% prior to chemical etching) from indium(III) acetate and gallium(III) 2,4-pentanedionate in octadecene at 280°C within minutes [43]. Even higher (approximately 47%) QY InP nanocrystals have been prepared by using MW treatment in the presence of a fluorinated ionic liquid, 1-hexyl-3-methyl-imidazolium tetrafluoroborate. This allowed for a very convenient approach, both for growing and *in situ* etching of the InP nanocrystals without the need for post HF treatment [46]. In another work the MW assisted synthesis of CdSe QDs from $Cd(OAc)_2$ and Se precursors was achieved using a Mars 5 X CEM system in cyclohexanone at 145°C [47]. The CdSe particle size was varied from 2 to 6 nm and was controlled by the time of the microwave treatment.

In fact, even using a domestic microwave oven and inexpensive precursors, highly luminescent CdSe/ZnS core/shell nanocrystals can be easily prepared [48] illustrating how microwave heating is a highly promising, significantly less dangerous and easily accessible method for the synthesis of QDs.

1.2.4 Transfer of QDs from Organic to Aqueous Phase

Synthesis in organic solvents gives QDs with a high percentage of luminescence quantum efficiencies at room temperature (up to 85%). In the most of cases, even though these organometallically prepared nanocrystals possess high crystallinity, narrow size distributions and high fluorescence QYs, their further use in bioconjugation or imaging is hampered by their lack of solubility in aqueous solutions. Therefore, for potential biological applications of QDs, phase transferring from organic to aqueous has been utilised to produce water-based colloidal solutions.

Several strategies have been pursued to solubilise high-quality hydrophobic QDs in aqueous solvents. Initial developments in this area were conducted by the Quantum Dot Corporation (now Invitrogen). They developed and patented [19 issued patents, including pioneering patent: U.S. Patent No. 5,990,479] aqueous CdSe nanocrystals for bioimaging by their encapsulation in a silica shell and their conjugation to various biomolecules (Fig. 1.5).

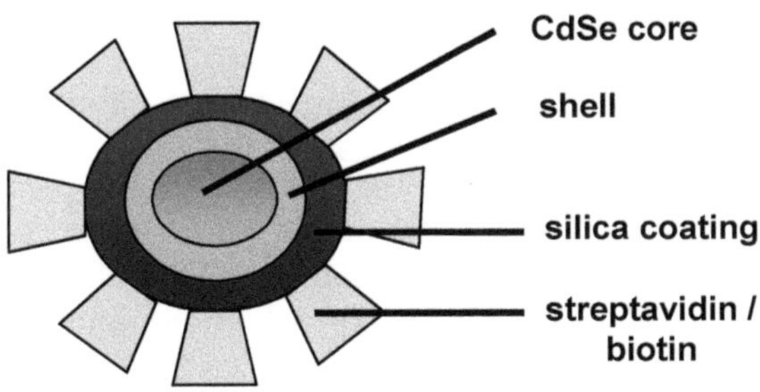

Figure 1.5 Graphical representation of a nanocrystal produced by the Invitrogen.

The silanization method which involves the growth of a silica shell around QD nanocrystals has been initially reported by Bruchez *et al.* [49]. In this approach (Fig. 1.6) [50] the TOPO-capped CdSe/ZnS QDs are initially treated with mercaptopropyltris(methyloxy) silane which is replaces the TOPO molecules on the surface after basification. During this process the methoxysilane groups (Si–OCH_3) are hydrolyzed into silanol groups (Si–OH). Then, the particles can be reacted with fresh silane precursors containing an appropriate functional group (F = –SH, –NH_2, etc.), which enables further surface modifications of nanoparticles. These core–shell QD structures have a greater stability in water and biological buffers as well as good biocompatibility.

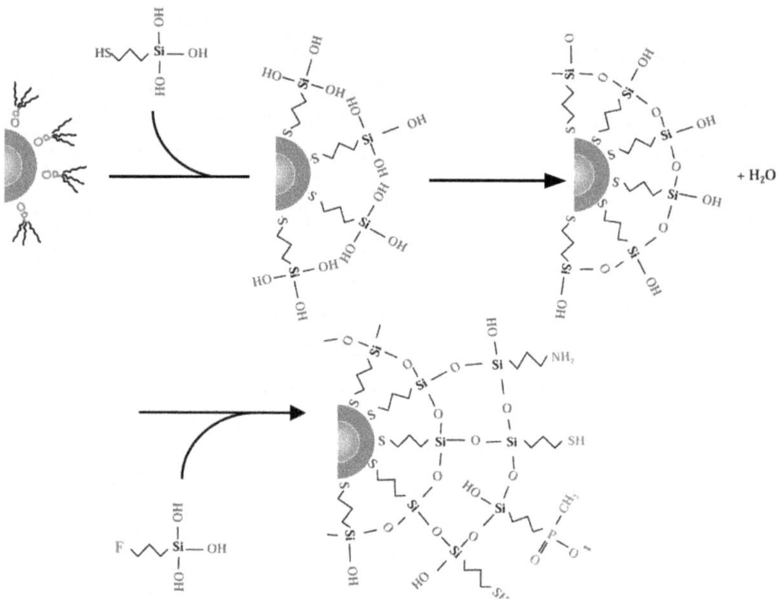

Figure 1.6 Sketch of the silanization method. Reproduced from 50.

Another approach involves the transfer of TOPO-capped QDs to water by exchanging the TOPO molecules for other bi-functional molecules, which contain a functionality that is able to bind to the QD surface (e.g., thiols, phosphines) and a hydrophilic group (e.g., carboxylic acid, or amino groups) This process takes advantage of the fact that the coordinating molecules such as TOPO are not covalently bound and can be easily replaced, for example, by other thiol ligands. Reiss and co-workers [33] reported the use of mercaptoundecanoic acid (MUA) for transferring CdSe nanocrystals into water. They found that uncapped CdSe cores exhibited a dramatic decrease in their luminescent properties upon transferring to water; however, those that were capped with a shell of ZnS displayed no adverse effects. Chan and Nie [51] utilised thioglycolic acid (TGA) to transfer the CdSe nanocrystals into water with the goal of attaching a protein to the nanocrystal surface. They were then incubated with HeLa cells and the experiments performed demonstrated the nanocrystals aggregated with antibody recognition of immunoglobulin G. Mitchell *et al.* [52] used 3-mercaptopropionic acid (MPA) to initially passivate the nanocrystal surface and act as a pH trigger for controlling the water solubility. Alkylthiol-capped oligonucleotides were subsequently immobilised on the surface and hybridised to a complementary strand.

In another work Dubertret *et al.* [53] reported the encapsulation of QDs into the hydrophobic core of phospolipid block-copolymer micelles This approach allowed for the production of stable, non-toxic, cell autonomous, and slow to photobleach QD nanocomposites.

Similarly, high-molecular-mass amphiphilic triblock copolymers consisting of a poly(butylacrylate) part (hydrophobic), a poly(ethylacrylate) part (as hydrophobic), and a poly(methacrylic acid) part (as hydrophilic) have been utilised to encapsulate the QDs to improve their water solubility and biocompatibility (Fig. 1.7) [54, 55].

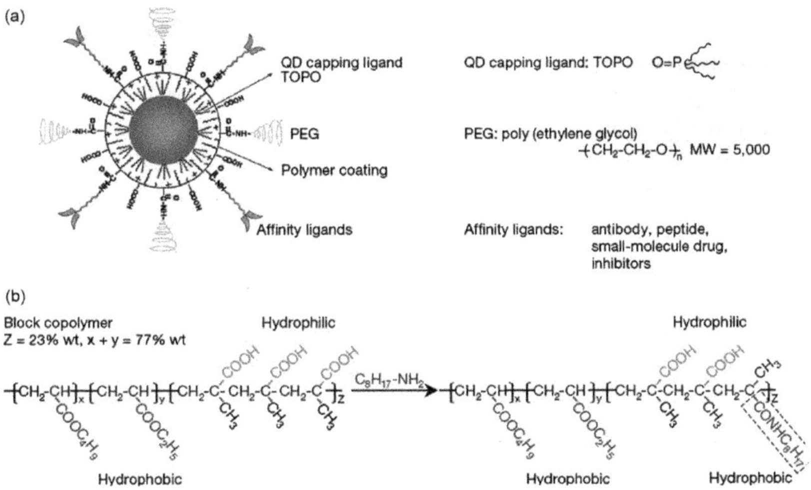

Figure 1.7 (a) A schematic illustration of a QD coated with a triblock copolymer shown in (b). Each of the blocks of the copolymer is designed to perform a specific function, e.g. protection *in vivo*, ligand targeting and improvement of biocompatibility and circulation. The polymer spontaneously encapsulates the TOPO-coated nanoparticles through strong hydrophobic interactions. Subsequently, PEG chains or biologically active molecules are grafted onto the polymeric shell. Reproduced from 54.

1.3 Direct Aqueous Syntheses of QDs

Direct aqueous synthesis by using short-chain thiols as stabilizing agents offers an excellent alternative to the previously mentioned synthetic approaches, which employ organometallic reagents and high-boiling-point organic solvents.

Because of the success of organometallically synthesised CdSe, explorations into their production under aqueous conditions have been sparse [56–58] and frequently yielded polydisperse nanocrystals exhibiting trap emission which overlaps with the intrinsic luminescence. Rogach *et al.* [56] have demonstrated, as an expansion of the wet-chemical route for the production of II–VI semiconductor nanocrystals, that CdSe can be synthesised using mercaptothiols (2-mercaptoethanol and 1-thioglycerol) and mercapto acids (thioglycolic acid and thiolactic acid) as stabilisers. They found that smaller nanocrystals (1.2–2.2 nm in diameter) could be produced with the thiols while larger nanocrystals (2.1–3.2 nm) were produced using thioacids. The same group have also demonstrated the production of HgTe [59] using $Hg(ClO_2)_4$ and bubbling H_2Te through at basic pH values in the presence of 1-thioglycerol.

The production of non-aqueous CdTe nanocrystals has had a relatively limited development [60–63], while its aqueous synthesis has expanded rapidly over the last few years. Precursors such as NaHTe [27, 64–66] and H_2Te [67–71] are most commonly used but variations such as $(NH_4)_2Te$ [72] and Na_2TeO_3 [73] have been investigated recently. Highly luminescent CdTe nanocrystals have also been synthesised using microwave irradiation [66]. The aqueous-based preparation of mercaptoethanol and thioglycerol-capped CdTe nanocrystals, was first investigated by Rogach and co-workers [74]. In a typical standard CdTe synthesis, a basic solution (adjusted with NaOH) of $Cd(ClO_4)_2$ and an appropriate amount of the thiol stabiliser is treated by H_2Te gas, which is generated by the chemical decomposition of Al_2Te_3 powder by concentrated sulphuric acid (see Scheme 1.1). This process initially results in non-luminescent CdTe-based clusters, which are then heated under reflux at 100°C for several hours to afford a crude solution of CdTe QDs [75]. The size of the QDs can be tuned by altering the reflux time with elongated reflux times resulting in larger QDs. To synthesise even larger QDs (diameter >6 nm), an additional cadmium source and stabiliser coupled with a further injection of H_2Te is required. This results in the formation of new nano-clusters, which can then add onto the already formed particles *via* Ostwald ripening.

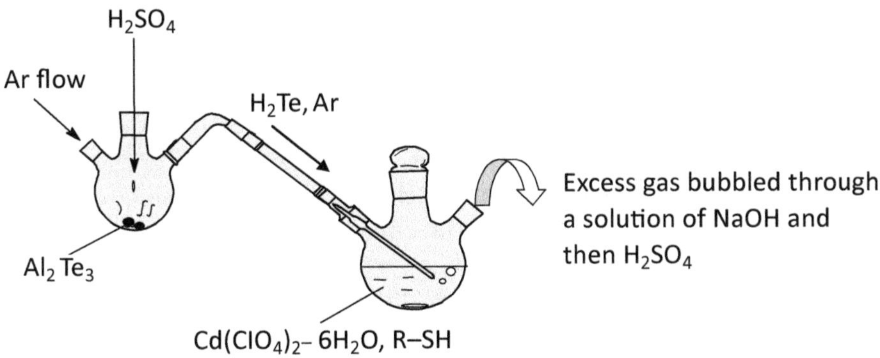

$$Cd(ClO_4)_2 + H_2Te \xrightarrow{\text{HS-R}} Cd\text{-}(SR)_xTe_y + 2HClO_4$$

$$Cd\text{-}(SR)_xTe_y \xrightarrow{100^\circ C} \text{CdTe-nanocrystals}$$

Scheme 1.1 Schematic illustration of the setup used to produce water-soluble CdTe QDs. Al_2Te_3 is decomposed by the dropwise addition of H_2SO_4 and argon carries the evolved H_2Te gas through a solution of $Cd(ClO_4)_2{\cdot}6H_2O$ and the desired stabilising thiol. This solution is then heated under reflux.

The same group also investigated the production of HgTe [59] and CdHgTe core–shell [76] nanocrystals soluble in an aqueous environment. Kapitonov *et al.* [64] were among the first to elucidate the trap states at lower temperatures and carried out lifetime studies for CdTe nanocrystals in water. Gaponik *et al.* [67] were among the first to use H_2Te (however, this is a very toxic gas [72, 77]) as a tellurium source and carried out thorough investigations into the properties, stability and photoetching of these nanocrystals. They found that thiols acted as efficient passivants due to their incorporation into a mixed CdTe(S) phase on the surface. This phase was further investigated by our group using Raman spectroscopy [69].

The utilisation of microwave heating has also a significant impact on the aqueous synthesis of QDs. For example, microwave techniques enabled a convenient preparation of alloyed QDs

(CdSe–CdS) in aqueous solution. In this work, CdSe seeds were first produced by the reaction of NaHSe and Cd^{2+} followed by microwave heating using 3-mercaptopropionic acid as sulphide source to generate the CdSe–CdS QDs [78]. These alloyed QDs demonstrated good photoluminescent properties and QYs up to 25%. Even higher QYs (between 40 and 80%) were achieved for CdTe/CdS/ZnS core/shell/shell QD structures, which have been prepared in the aqueous phase by MW heating [79]. In this case, the growth of an intermediate CdS layer (to minimise the lattice mismatch between CdTe and ZnS) followed by ZnS allowed a smooth change of lattice spacing from the emitting CdTe core to the protecting ZnS shell, which reduces non-radiative surface defects and improved the QY. Moreover, these core–shell structures not only have a high QYs but also demonstrate good aqueous dispersibility, an excellent photostability and very favourable *in vitro* biocompatibility [79].

Finally, the utilisation of an aqueous microwave approach enabled the synthesis of strong defect emitting chiral *D*- and *L*-penicillamine-capped CdS QDs, which have shown almost identical mirror image responses in their circular dichroism spectra in the range of 200–390 nm (Fig. 1.8). This new type of QDs could find very important applications, including their use as fluorescent chemical and biochemical chirality sensors and molecular recognition nanodevices [80].

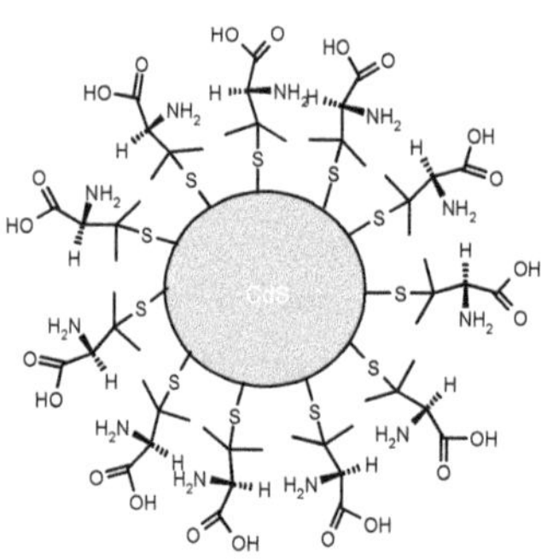

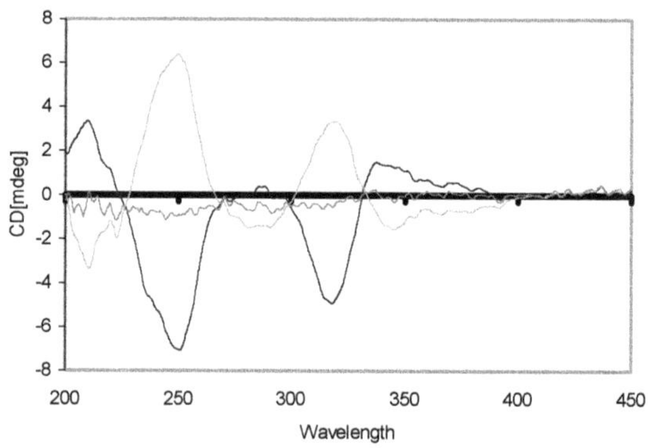

Figure 1.8 Left: schematic presentation of the *D*-penicillamine-stabilised CdS nanoparticle. Right: CD scans of the *D*- (Blue)-, *L*- (Green)- and *Rac*-penicillamine (Red)-stabilised CdS QDs (θ deg = δA/32.98). Reproduced from 80.

1.4 Micellar Synthetic Approaches

Micellar techniques offer very interesting soft chemistry routes for synthesis of new QD materials. In this approach micelles can serve as nano-sized reactors, where different reagents can be mixed and processed together. The reverse (or inverse) micelle technique is particularly good for tailoring particle size and distribution. Reverse micelles are also excellent candidates for producing QDs of the same shapes as their nuclei [81]. Normally reverse micelles are small nano-sized water droplets, which are stabilized by a surfactant (e.g. sodium salt of bis(2-ethylhexyl)sulfosuccinate) in a non-polar medium (e.g. oil). Thus, the water structure of the droplet is one of the main parameters, which allows control over the nanocrystal size. Pileni *et al.* have demonstrated that the reverse micelle method can be successfully utilised to prepare CdTe QDs, but the researchers have found that this technique cannot be used to produce $Cd_{1\pm y}Mn_yTe$, $Zn_{1\pm y}Mn_yTe$, or ZnTe structures that are thermodynamically stable in the bulk phase [82]. The same group has also reported the synthesis of unusual triangular shaped CdS QDs from a cadmium bis(ethyl-2-hexyl)sulfosuccinate precursor in isooctane/water reverse micellar solution [83]. There have also been several further publications on the preparation of CdS QDs using the reverse micelle method [84–88]. For example, CdS nanocrystals were synthesized in reverse micelles of water/cetyltrimethyl ammonium bromide $(CTAB)^{+4}$-penten-1-ol/benzene. It was found that the micellar environment has a significant impact on the dynamics during the growth of the QDs. Initially the micelles, which contained only Cd^{2+} and S^{2-} precursors, the growth was continuous under relatively fast processes with timescales of several tens of minutes. However, when monomer and cross-linker molecules were loaded in the micelles, Ostwald ripening growth was observed under slower processes with increased timescales of several hours [85]. CTAB was also used as a surfactant to prepare the nanodispersions of CdS and HgS as well their core–shell products and composites by varied modes of precursor addition in micellar solution [86]. In another work CdS nanorods have been prepared *via* arrested precipitation from cadmium and sulphur ions in a micellar solution [88].

Other QD systems, which have been effectively produced by the reverse micellar technique include CdSe [87, 89] PbS, and [90] ZnS [91] as well as $Zn_{1-x}Cd_xS$ alloy-based QDs [92]. A new approach involving self-assembling block copolymer micellar templates and gas-phase reactions was also utilised to prepare arrays of monodisperse III-nitrides (GaN) nanocrystals in the size range of 1–5 nm with uniform spacing's between the nanocrystals [93].

In addition, reverse micelles facilitated the coating of the QDs with a very thin silica layer within the same reverse micellar system. For example, the controlled hydrolysis and condensation of tetraethyl orthosilicate (TEOS), using an ammonium hydroxide catalyst and the water-in-oil microemulsion resulted in the production of highly luminescent and photostable silica coated CdS:Mn/ZnS core/shell QDs [94].

Finally it was recently reported that reverse-micelle synthetic approach has been used to produce new QD-based barcode systems [95]. In this work, the encoding elements, Zn^{2+}, Cd^{2+} and Pb^{2+}, were integrated within a single QD by mixing the corresponding micellar solutions and the S^{2-} containing micellar solution (Fig. 1.9). This process resulted in new QD barcodes, which were subsequently decoded by a voltammetric stripping technique. The researchers have shown that these new barcodes can be potentially utilised as tags in sensitive immunoassays of a cancer marker, carcinoembryonic antigen.

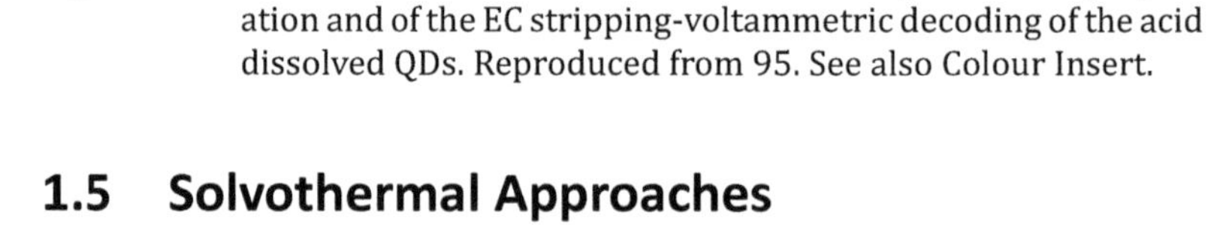

Figure 1.9 Illustration of the reverse-micelle-based QD barcode preparation and of the EC stripping-voltammetric decoding of the acid dissolved QDs. Reproduced from 95. See also Colour Insert.

1.5 Solvothermal Approaches

As discussed above (see Section 1.2), high-temperatures synthesis of QDs in organic solvents have several drawbacks. One of them is the use of high-boiling-point solvents, which makes the synthesis quite expensive and not energy efficient. The solvothermal technique offers a clever alternative approaches to address the problems above. In this approach, the solvent can be heated in a sealed vessel (autoclave or calorimetric bomb). This enables the autogenous pressure to exceed the ambient pressure and raises the effective boiling point of the solvent. Therefore, this method allows the use of a cheaper solvent with a lower boiling point for the synthesis of the desired nanomaterial [96]. In addition, the solvent can be heated to a temperature above of its critical point enabling it to dissolve frequently insoluble materials at normal conditions consequently speeding up the reaction. In the case of the solvent being water, the technique is often referred to as hydrothermal method.

One of the early reports in this field was on the synthesis of the nanocrystalline III–V semiconductor InAs from $InCl_3$, $AsCl_3$ and Zn powder in xylene by the solvothermal treatment at T = 150°C [97]. A similar approach also allowed the synthesis of InP, and GaN QDs using convenient and inexpensive inorganic precursors [98, 99]. Even elemental precursors can be used for the synthesis of QDs by the solvothermal technique. For example, ZnSe nanoparticles have been prepared by solovothermal heating of the corresponding elements (Zn and Se) in pyridine at T = 180°C [100]. Also ZnSe and CdSe QDs have been prepared from the elements in water using the low temperature hydrothermal method [101].

A range of various CdS and CdSe QDs have been prepared using redox and chemical precipitation reactions under solvothermal conditions. For example, CdS QDs of different morphologies have been prepared in different solvents from cadmium nitrates or sulphates using thiourea as the S source [102]. Small-size (3 nm) CdSe nanoparticles have also been synthesised from cadmium stearate and Se powder by a solvothermal reaction in toluene using tetralin (tetrahydronaphthalene) as a reducing agent [103].

In other work, highly crystalline wurtzite CdSe nanorods and zinc blend CdSe nanoparticles have been produced by the solvothermal treatment of cadmium nitrate, hydrazine hydrate and Se in ethylenediamine at 140°C [104, 105]. CdSe nanorods of various length and diameter have also been prepared using similar solvothermal techniques in different solvents [106–108].

Overall, the solvothermal approach is quite versatile and enables the generation of nanomaterials of various shapes and sizes. However, this technique frequently gives very low yields and products with poor uniformity in both size and morphology. There are also some issues with the reproducibility of the synthesis and will require far more research to enable this technique to be widely utilised for the synthesis of high-quality QDs.

1.6 QD Luminescence Improvement

Across all of the synthetic techniques that have been discussed so far, one common goal runs throughout. Each one aims to produce highly luminescent stable QDs. Consequently, the nature of the QD surface, the effectiveness of the passivating ligands and the removal of non-radiative defect sites is paramount. As such, various techniques and strategies have been employed to improve the emission properties of the QDs and will be further discussed.

1.6.1 Surface Passivation and Core–Shell Structure Formation

One very important characteristic of QDs is the abundance of atoms that reside on their surface [17, 109, 110]. QDs possess an extremely high surface-to-volume ratio, and as such, the nature of the surface becomes critical and its passivation or coordination can lead to the formation of surface trap states that can strongly influence the recombination of the charged carriers. Semiconductor nanocrystals typically contain quite a high density of defect states and since

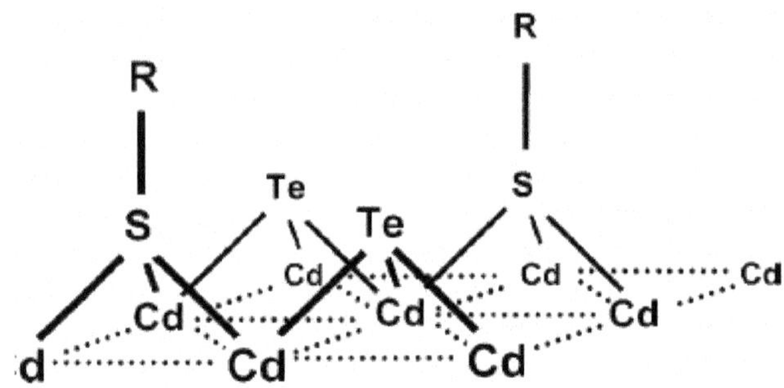

Figure 1.10 Graphical illustration of the mixed occupancy on the surface of a CdTe QD. Note: when surface Te sites dominate, this results in a high quantity of surface dangling bonds and lower luminescence efficiencies. The occupation of these three-fold coordinate sites with a thiol molecule results in increased stabilisation and the removal of surface dangling bonds (passivation). This also leads to the formation of a core–shell-like structure of Cd and S, which acts like a potential well giving a better confinement of the excitons and increased QYs for the nanocrystals. Reproduced from 70.

a large majority of the atoms reside on the surface, so do many of the defect sites (Fig. 1.10). It is important that we understand the processes occurring at these interfaces as the synthetic procedures employed for the production of QDs and the capping agents used, directly influence the nature of these defects.

Upon optical excitation, an electron is promoted to the CB and radiation occurs upon its recombination with the hole left behind in the VB (also termed radiative recombination). Recombination may occur between the bands, surface states or crystal defect sites which reside at energy levels within the band gap (Fig. 1.11). Energy loss occurs *via* the creation of phonons in the lattices which enable the free carriers to be trapped at those recombination centres, which may be radiative or non-radiative [111].

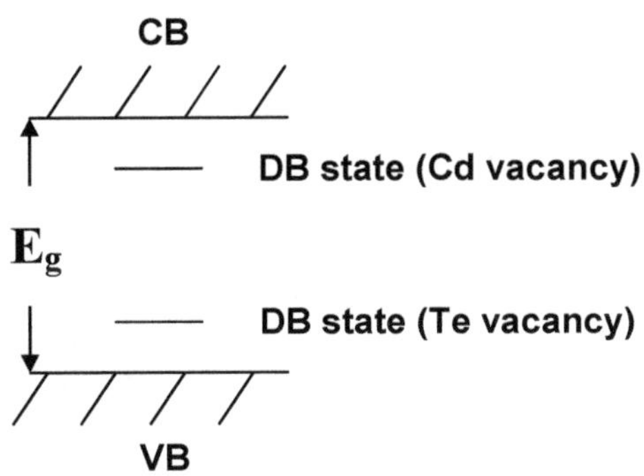

Figure 1.11 Illustration of the location of the anion (Te) site and the cation (Cd) defect sites within the band gap (E_g). Reproduced from 109.

In general, a high-luminescence QY results from predominantly intrinsic recombination, while a low QY is considered a result of the defect states located within the band gap acting as traps for the photogenerated carriers. These defect states provide alternate routes for the rapid (<1 ns) recombination of the exciton, resulting in radiationless transitions [112]. Borchert *et al.* [70] have found there were less Te atoms on the surface of highly luminescent CdTe QDs with respect to those with low luminescence. The removal of the anion (Te) site passivation leads to the formation of a dangling bond (DB) state just above the valence band, which acts as a trap for photogenerated holes. Conversely, the removal of the cation

(Cd) site passivation leads to an electron (e^-) trap just below the CB (Fig. 1.11) [109].

As such, the QD surface can be viewed as an ensemble of dangling and terminal bonds [22, 113–115] as well as vacancies, dislocations, screw and other defects which are all in various forms of coordination with the surrounding solvent and any associated ionic or adsorbed entities [110, 116].

Theoretical treatments indicate that the efficiency of the passivation provided by the surface ligands depends strongly on the surface structure and the nature of the surface states of the nanocrystals themselves [115]. To ensure a reduction in the number of surface dangling bonds and increase radiative recombination, alterations can be made to the stoichiometric ratios of the reagents [117].

The ligand dynamics depend highly on the reaction temperature and the ligand–ligand interactions greatly affect the surface quality [115, 118]. If the surface ligands can provide a good electronic passivation for the surface states of the nanocrystals, a high QY should be expected. However, due to both steric and electronic interactions, surface binding of a ligand may be hindered by the presence of another ligand. Investigations have shown that increased QYs can, in fact, be obtained by decreasing the ligand-to-metal ratio [70, 116].

While high QY values have been achieved, there are problems with this reduction in quantitative precursor amounts, primarily due to the formation of a variety of cadmium complexes in solution. The main cadmium complexes that play the significant role in determining the QY in solution are CdL_2^{2-}, CdL_3^{4-} and CdL (L = thiol stabiliser) [70, 117]. Each complex dominates depending on the ratio of cadmium to thiol used (Fig. 1.12).

At higher ligand-to-metal ratios (>2.5 : 1), $Cd(L)_3^{4-}$ dominates while at ratios of (2 : 1) CdL_2^{2-} becomes prominent. These comp-lexes are completely soluble due to the carboxylic acid stabilising groups present and these solutions remain transparent throughout the entire reflux duration. However, this is not the case for lower ligand ratios, where the complex CdL now dominates. This complex, due to the presence of only one thiol stabiliser, is not as soluble as those previously highlighted and gives rise to a distinct turbidity during reflux with the formation of a white precipitate. Additionally, this precipitate may also be due to the production of cadmium hydroxide ($Cd(OH)_2$). This is formed as a white, insoluble powder when sodium or potassium hydroxide is added to a solution of a cadmium salt [119].

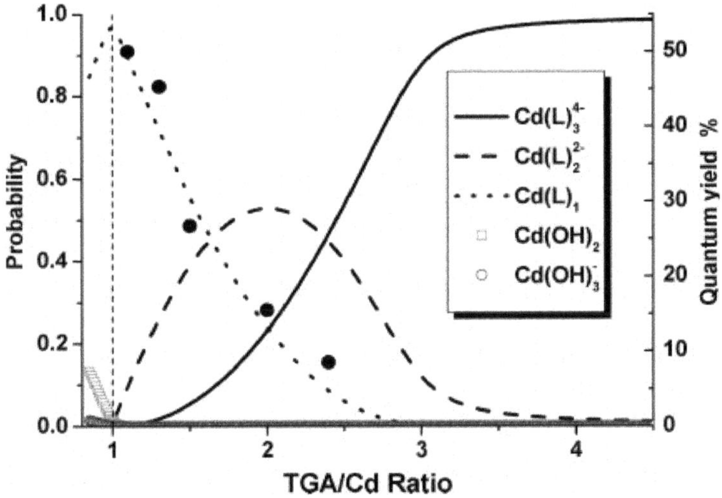

Figure 1.12 Graph illustrating the distribution of the various cadmium complexes in solution and the QY of the QDs after 20 hours of synthesis depending on the TGA/Cd ratio used. Adapted from literature. Reproduced from 117.

$$Cd^{2+} + 2OH^- \rightarrow Cd(OH)_2 \downarrow$$

The overall impact is that this precipitate must be filtered from the reaction mixture before further purification is undertaken. This results in a decrease of the overall cadmium and thiol concentrations in solution, giving rise to lower reaction yields and QD solutions which are less concentrated than those synthesised at higher ratios but with the benefit of higher QYs.

The CdL monomer can also release excess Cd in solution during the growth process, which acts to increase the amount of $Cd(OH)_2$. Nevertheless, for the production of large quantities of CdTe QDs, this reduced ratio process requires large amounts of cadmium in solution leading to a more costly procedure.

The capping of the nanocrystals by a wider band gap material (such as CdS, ZnS) also provides another route to increased luminescence intensity and QY. This generally occurs for QDs synthesised in organic solvents and numerous reports have indicated that an epitaxial layer of an inorganic passivant significantly reduces the number of non-radiative centres by efficiently capping the unsaturated surface atoms [9, 120–122].

The occupation of the vacant surface sites by, in this case, sulphur atoms with an additional bond to the organic stabiliser gives a more favourable passivation. When the Te vacancies are occupied by sulphur, the dangling bonds are removed and a core–shell-like structure is formed which acts as a potential well, enabling better confinement of the photogenerated carriers. This shell of material can be grown around the core inorganically in an organic medium but can also, in the case of thiol-stabilised, water-soluble QDs, be generated by the decomposition of the thiol capping reagent [69] and the incorporation of sulphur into the surface structure. In fact, Zhang *et al.* [123] and Gaponik *et al.* [67] have found that effective surface passivation by the formation of Cd–thiol complexes [27] was the origin of high-luminescence QYs.

1.6.2 Photoetching of QDs

The photoetching of the QD solutions has also been widely investigated as another method for the improvement in QYs [67, 124–126]. This process is quite an important route towards the production of highly luminescent nanocrystals and opens up possibilities for their use in photolithographic processes [127]. Photoetched QDs also demonstrated a significant potential as probes for specific *in vitro* cellular labelling [128].

The illumination procedure is dependent on a number of conditions; excitation wavelength, lamp power, nitrogen or oxygen environment and duration of irradiation. Also, the spectral properties of the QDs must be monitored as any shifts in emission or absorption peaks may indicate particle growth (due to the incorporation of excess reagents into an epitaxial shell) or sample degradation. Several theories have been proposed to explain the observed phenomena. The presence of oxygen has been found to play an intricate role in the luminescence improvement. Wang *et al.* [129] found that an approximate 10% increase was noted for CdSe QDs when irradiation was conducted in the presence of oxygen as opposed to 1.5% without. They suggest that a reconstruction or annealing of the surface takes place to remove defects and this is further confirmed as photoactivation was often accompanied by photocorrosion, as given by Scheme 1.2 and shown in Fig. 1.13.

$$CdSe + O_2 \xrightarrow{h\nu} Cd^{2+} + SeO_2$$

Scheme 1.2 Photocorrosion of CdSe to give free Cd^{2+} and SeO_2.

They conclude that the blue shift in the emission spectra that accompanies illumination is due to the removal of mid-gap states and structural features from the surface [130].

The annealing of these defects (such as dangling bonds associated with unsaturated Te atoms) should be higher for those QDs possessing a large quantity of defects [129, 28]. These and other defect states are broader and less pronounced and appear as an independent feature which is red shifted compared to that of the bandgap emission [131]. At low temperature, these states can be clearly resolved and identified [64], however, at room temperature, they appear to cumulatively add to the intrinsic emission resulting in spectral broadening, increased FWHM values and a decreased symmetry of the emission spectrum.

The gradual erosion of the topological features arises from the trapping of the charged carriers in surface states. Here the excitons activate reactions towards the oxidation of O_2 to O_2^- and Se to SeO_2, resulting in the removal of atomic moieties from the surface and giving rise to a shift in the emission spectrum to lower wavenumbers. This surface curing results in the reconstruction of the stabiliser shell and annealing of the surface defects (Fig. 1.13) [28].

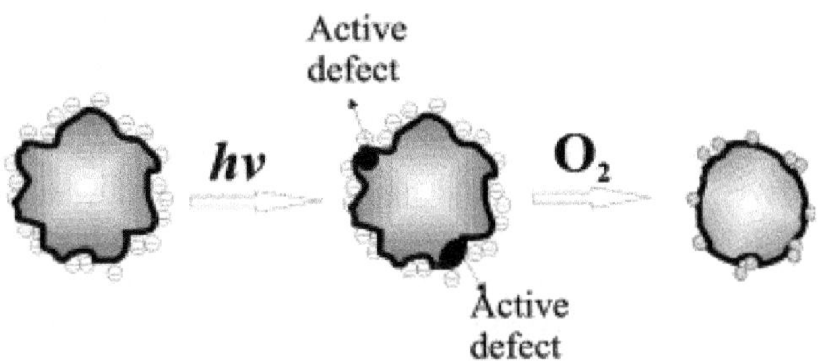

Figure 1.13 Illustration of the trapping of the charged carriers in "rough" states where photooxidation eliminates the atomic scale imperfections. This reduces the size of the crystallite and enhances the QY. Reproduced from 129.

An increase in QD luminescence can also be achieved by capping the surface with a material possessing a larger bandgap such as

CdS or ZnS [18, 19, 33]; however, this is typically accompanied by an increase in overall size. This layer effectively caps the defects present on the surface, thus reducing the number of non-radiative channels and increasing the overall QY. For different types of QDs, different shell compounds are created. Bol *et al.* [132] have shown that irradiation and subsequent oxidation of ZnS:Mn_2^+ crystallites results in the production of $ZnSO_4$ and $Zn(OH)_2$, which serve as a passivation layer. Dunstan *et al.* [111] have shown that for ZnS QDs the decomposition products are SO_4^{2-}, S, Zn and SO_3^{2-} and these species act as individual recombination centres at the surface. While these processes result in the formation of new layers at the QD surface, it has also been postulated that the decomposition of thiol capping molecules results in increased QY values [67, 69, 133]. Zhang *et al.* [133] used XPS (X-ray photoelectron spectroscopy) to experimentally observe that Cd–thiol complexes on the crystal surface provide the crucial chemical passivation responsible for high luminescence efficiency. Also, the high stability and QY corresponded to the QDs with the greatest Cd–thiol surface coverage, as S was found to occupy Te sites in the lattice. Holes have also been shown to oxidize sulphide anions [134] (Scheme 1.3), creating elemental sulphur which can add to and cap the nanocrystal surface.

$$H^+ + S^{2-} \rightarrow S^-$$
$$S^- + O_2 \rightarrow S + O^{2-}$$

Scheme 1.3 Mechanism of the oxidation of sulphur by photogenerated holes as proposed by Henglein. Reproduced from 133.

The composition of the QDs themselves may also contribute to any effects witnessed. It has been shown that different QDs are capable of producing free radicals in solution (CdS — hydroxyl and superoxide; CdSe — hydroxyl; CdSe/ZnS — no radicals) [135].

The initiation of corrosion by the photoinduced carriers can result in the generation of the elemental components from the QDs [136]. These photocorrosion products may cumulatively add onto the surface and the formation of an oxide layer can result in increased passivation [132, 137, 138]. The generation of these elemental components under these conditions is of particular concern for the use of QDs in bioimaging and cellular investigations. As such, the capping ligands play a significant role towards the efficient passivation of the surface by successfully reducing the number of adsorbed oxygen-containing species available for radical formation.

1.6.3 Role of Capping Ligands

Coupled to this oxidation of the QDs, the stability of the capping ligands plays an important role both in the preparation of the QDs and in the control of their optical properties. This is further emphasised by the fact that before purification from the crude parent solution, the QDs can be stable for years. However, after size-selective precipitation, the surface coverage can be compromised due to the removal of thiols by the addition of a non-solvent. On a positive note, this instability can be utilised to generate new nanostructures such as tetrapods or wires [139–142]. QD instability can manifest itself in three distinct processes [125]: the photocatalytic oxidation of the stabilising ligands (generating disulphides), the photooxidation of the nanocrystal itself (as shown previously) and finally precipitation.

Owing to their large surface to volume ratios and the quantity of atoms present on the QD surface, the capping ligands act to both sterically prevent each individual particle from aggregating and to neutralise surface dangling bonds due to unpassivated atoms [110, 113, 143]. The activation and subsequent decomposition of these capping agents gives one explanation as to why irradiation times can vary from sample to sample and why the decomposition of the QDs and their subsequent aggregation and precipitation from the solution arises. Fischer and Henglein [144] have shown that thiolate (SR^-)-stabilised CdS nanocrystals photodissolve in the presence of oxygen; however, they are also sensitive to irradiation in its absence. They proposed the mechanism shown in Scheme 1.4, which illustrates the various processes initiated by the absorption of light.

Scheme 1.4 shows the mechanism proposed to explain the removal of the thiol capping group from the surface of CdS nanocrystals. The thiol ligands can be gradually photoxidised to disulphides, with a net result of a loss in the number of stabilising groups. However, at a certain threshold of illumination time, the QDs will lose a critical amount of their capping groups and due to this thiol loss, their stabilisation is compromised and they are no longer able to remain in solution.

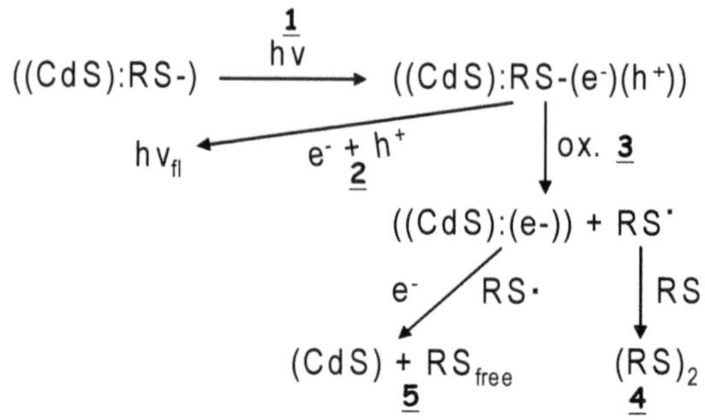

Scheme 1.4 Mechanism proposed by Fischer and Henglein [144] which describes the photocatalytic removal of the surface capping thiol molecules from CdS nanocrystals. **1:** light absorption generates an exciton (e^-/h^+ pair), **2:** exciton can recombine radiatively (fluorescence, fl) or non-radiatively (main reaction), **3:** oxidation of thiolate anion to form a thiyl radical which is released from the nanocrystal. **4:** thiyl radical can dimerise to form the disulphide. **5:** electron can pick up another one to form CdS and free thiol.

Another contributing factor is pH, which also affect greatly the QD and ligand stability and consequently the QY. It has been shown that while the QDs are synthesised at basic (~11) pH values for carboxylic acid capping agents, a reduction in pH to approximately 4.5 can generate an improvement in the QY. Any further lowering of the pH results in decomposition [145].

Gao *et al.* [145] determined that the luminescence efficiency depends greatly on the pH of the solution. They witnessed a dramatic five-fold increase in QY upon decreasing the pH. They attribute this behaviour to the structural reorganisation of the QD surface, due to the formation of a Cd–TGA complex shell in acidic ranges. This complex, formed from the hydrolysis of the stabiliser, essentially acts as an additional capping reagent and has been explored by various other groups [28, 67, 123]. Aldana *et al.* [146] have also analysed pH alterations and found that the dissociation of ligands from the surface of the QDs was the main cause for their precipitation in relatively acidic solutions. However, they also state that the equilibrium pH values for this dissociation process were highly dependent on the size of the QDs.

The interaction due to the different functionalities of the various capping groups also plays a role in nanocrystal stabilisation and luminescence QY. By analysing the crystal structure of $Cd(SCH_2COOCH_2CH_3)_2$, Dance *et al.* [147] proved the existence of secondary coordination behaviour between the carbonyl oxygen from ($SCH_2COOCH_2CH_3$) and the primary thiol combining cadmium. This interaction, depending on the chain length of the thiol molecule, gives rise to a stable five- or six-membered ring formation (Fig. 1.14). An enhancement of the QY was also observed by Zhang *et al.* [123] for 3-mercaptopropionic acid-stabilised CdTe upon reduction of the pH to 6. They attributed the dependent luminescence behaviours of the thiol-stabilised QDs to the protonation of carboxylate groups from particle surface binding thiols. They also investigated, with the use of carboxylic acid rich polymers, the effects that may occur when the interference from surface binding thiols was removed (Fig. 1.14).

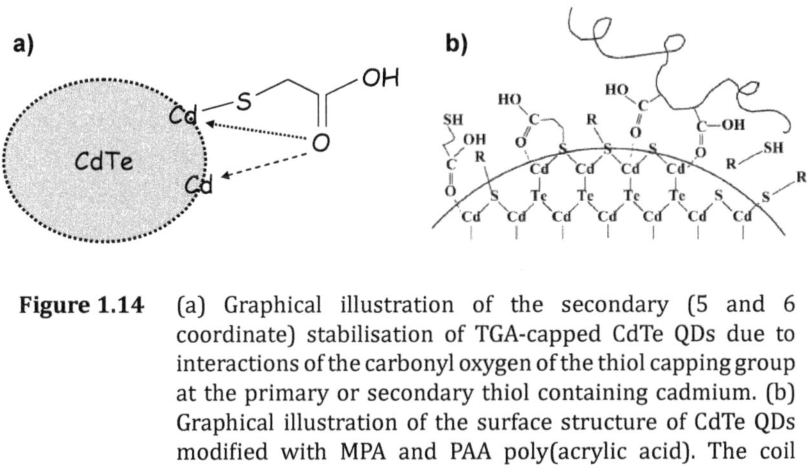

Figure 1.14 (a) Graphical illustration of the secondary (5 and 6 coordinate) stabilisation of TGA-capped CdTe QDs due to interactions of the carbonyl oxygen of the thiol capping group at the primary or secondary thiol containing cadmium. (b) Graphical illustration of the surface structure of CdTe QDs modified with MPA and PAA poly(acrylic acid). The coil represents the PAA alkyl chain and R represents a carboxylic acid group. The coordination between the carbonyl oxygen and cadmium on the particle surface is represented as a dotted line. The naked sulphur species in the particle results from the decomposition of thiols which occurs in the acidic pH region. Reproduced from 123.

QY increases were demonstrated indicating that carboxylic acid moieties can indeed provide an additional stabilisation for the QD surface. This reinforces the theory that secondary interactions from external or internal (thiol capping group) can in fact lead to luminescence increases.

In another work, a mixture of gelatin polymer and TGA was used for *in situ* synthesis of CdTe QDs. It was found that the gelatin alters the nucleation and growth of the QDs and results in large increases in the resultant luminescence efficiencies of up to 1.5 times their original values for just TGA-capped QDs. This gelatine coating also resulted in a significant reduction in the *in vitro* toxicity of the QDs possibly opening up new opportunities for biomedical applications of new QD nanocomposites [148].

1.7 Further Functionalisation and Bioconjugation of QDs

Research into the binding of QDs to various biomolecules has seen a rapid expansion over the last 10 years as the search for stable, highly luminescent fluorophores as alternatives to traditional organic dyes and potential drug delivery agents has intensified.

The conjugation of QDs to biomolecules has been shown to reduce aggregation and improve stability within cellular regions [51, 149–151]. Conjugation strategies follow three major routes: (i) 1-Ethyl-3-(3-dimethylaminopropyl) carbodiimide (EDC) coupling enables cross conjugation between carboxylic acids and the primary amino functionalities of the biomolecule [149, 152]. However, this can lead to problems of cross-linking due to the variety of functionalities present within the protein or oligonucleotide which allow for non-specific and indiscriminate binding. (ii) The biomolecules can possess binding moieties themselves and can be directly attached to the surface, generally through thiol containing cysteine [21, 153] linkages or *via* the metal-affinity coordination of histidines [154]. (iii) Non-covalent binding involving the use of electrostatic self-assembly to conjugate nanocrystals to proteins, antibodies or intermediate molecules [155–157].

The labelling of cells can be distinctly separated into two categories: targeted *via* bio-conjugation and non-specific labelling. A key issue for the use of QDs in cell tracking is their delivery to the cytoplasm or organelles such as the nucleus or mitochondria. While there are various dyes available which can penetrate cell membranes easily (although they can adversely affect the cells) and label these cellular features (DAPI, Mitotracker, Hoescht), QD properties such as size and functionality may prevent crossing of the lipid bi-layer. As such, Derfus *et al.* [158] have explored several strategies for enhanced delivery of QDs into live cells. They found that QDs were sequestered by endocytosis into the endolysosomal compartments, which renders them unavailable for further assays, but microinjection delivered the nanocrystals as monodisperse entities. However, this requires specific and individual manipulation of the cells and is extremely time consuming.

Different types of functionalisation (primary antibodies [159], DNA [152], serotonin [160], biotin-avidin [21], avidin [157], chitosan [161] and peptides [153]) have been utilised to explore how QDs target specific cell proteins or receptors. Other moieties such as antibody (e.g. immunoglobulin G and streptavidin)-linked QDs were shown to label the Her2 breast cancer marker on the surface of cancer cells, stain actin and microtubule markers and detect antigens in the nucleus [162]. Also, QDs linked to DNA probes were reported as new nanosensors for the detection of low concentrations of DNA by FRET [152].

Clarke *et al.* [163] have investigated the conjugation of dopamine (an electron-donating molecule) to QDs in an effort to increase specific binding and recognition by cells that bear dopamine receptors. The conjugates display a redox-sensitive patterning, and oxidation or reduction results in particular localisation of the nanocrystal. Antioxidants were also shown to reduce the level of toxicity from any singlet oxygen produced upon irradiation.

Winter *et al.* [164] have attached CdS QDs to live neurons utilising both antibody and peptide recognition molecules, while CdSe/ZnS nanocrystals labelled with the protein transferrin underwent receptor-mediated endocytosis when cultured with HeLa cells. Those QDs labelled with immunomolecules were shown to recognize specific antibodies or antigens [51]. Neurons have further been illuminated by Dahan *et al.* [165], who showed that nanocrystal size controls penetration into neural synapses where glycine receptor tracking was possible over extended time durations when compared to organic molecules (20 minutes in this case).

Pathak *et al.* [166] have also conjugated 5′ aminated oligonucleotides to produce stable water-soluble CdSe/ZnS QDs. They found that these nanocrystals could be used for specific binding to cells containing Y chromosomes in an *in situ* hybridisation assay.

Autofluorescence from tissues due to the presence of low amounts of fluorophores (e.g. nicotinamide, flavins or collagen) or overlap of the QDs emission with that of the fluorophore can often impair the resolution of the fluorescence imaging [167]. As we have discovered ourselves, this overlap of emission wavelengths can result in a masking of the QDs fluorescence. This problem can be somewhat circumvented by using near infra-red (IR) excitation (window from 650–900 nm), which has been shown to penetrate deep into tissues: 10 cm of breast tissue and 4 cm of skull tissue [168]. As demonstrated by Weissleder and co-workers [168], near-IR light is appealing due to the low overlap with biological systems such as haemoglobin and water, which have their lowest absorption coefficients in this region (Fig. 1.15).

Typically, imaging in this area is conducted using cyanine dyes [168, 169] and has been investigated by Becker *et al.* [170], who imaged tumours displaying somatostatin receptors by using a cyanine fluorophore coupled to an analogue of somatostatin. Since QDs can have their emission tuned to a desired wavelength, it has been possible to prepare near-IR-emitting nanocrystals with imaging in mind. Such QDs have been synthesised and characterised but not biotested (although they can be transferred into an aqueous medium) by Harrison *et al.* (CdHgTe) [76], Bailey *et al.* (CdSeTe) [171], and Bakueva *et al.* (PbS) [172].

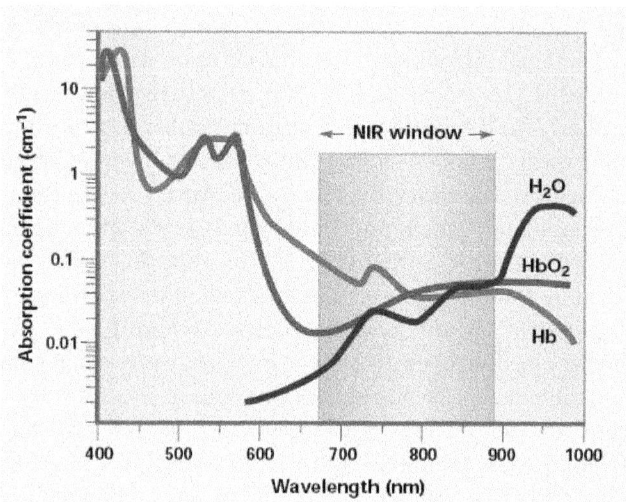

Figure 1.15 Near-IR window available for *in vivo* imaging illustrating the minimal light absorbed by haemoglobin and water in this region. Reproduced from 168. See also Colour Insert.

However, only a few reports into *in vivo* models have been demonstrated. Kim and co-workers have made use of CdSe, CdTe [173] and alloyed QDs such as $InAs_xP_{1-x}$ core/InP intermediate and CdSe shells [174] that emit in the near-IR (840–860 nm) for mapping sentinel lymph nodes in cancer surgery. Only 400 pmol of QDs were required and enabled the researchers to visualise 1 cm deep into the tissue. Levene *et al.* [175] have utilised near-IR emitting QDs for *in vivo* multiphoton microscopy of deep brain tissue in mice to depths of hundreds of micrometers, while capillaries have also been visualized by Larson *et al.* [176] and Akerman *et al.* [153], who conjugated peptides and targeted tumours in mice.

1.8 Concluding Remarks and Future Outlook

Over recent years, there have been significant advances in the development of efficient synthetic approaches for the production of various types of semiconductor nanocrystals. Quantum dots have a great potential for many applications ranging from photonics and photovoltaics to biosensing and biological imaging. Recent progress in optical detection and sensing suggests a great future for QD-based assays in biological imaging, particularly in fluorescent tracking experiments that could improve our understanding of cellular recognition events and could even help develop new techniques for assessing bio-distribution and even screens for future therapeutics. Another very promising area is the development of multimodal nanocomposites, which could simultaneously serve as contrast agents and photosensitizers for photodynamic therapy. With this in mind, intensive work will be required to scale up the existing approaches and to develop them towards enhanced reproducibility at the industrial scale. However, there are a number of issues and problems that need to be deciphered, addressed and overcome.

Despite all recent progress made and the obvious potential of QD-based nanomaterials, the vast majority of them can be applied only in basic research, or can remain of academic interest only. The potential toxicity of these materials (due to their inherent composition of toxic elements), can significantly limit their range of applications, particularly those congruent with biomedical uses. QD toxicity is still poorly very primitively studied and understood discipline; therefore a significant part of any future QD work must be focused on the investigation of QD induced toxicity and the biocompatibility of these QD-based materials. Modes on toxicity must be verified and explored and new synthetic procedures for reducing their risk factors must be developed. This will be crucial for further development of industrial scale synthesis of QDs and other relevant fields of technology.

Another problem is related to frequently poor stability of QD assays over the period of time under ambient conditions. New methods to improve QD stability in solution and bench life time must be explored. This may be particularly important with regards to their utilisation in electronic devices, with one eye on their safety and stability, and also on equipment decommissioning and waste, especially in the current climate of increased reuse and recycling.

Finally, the cost of QD production should be also minimised. This will require easily scalable techniques with an access to cheap non-hazard precursors, low-energy synthetic approaches, inexpensive eco-friendly solvents and low-cost processing technologies and facilities.

References

1. A. P. Alivisatos, *J. Phys. Chem.*, 1996, **100**, 13226–13239.
2. A. J. Nozik, F. Williams, M. T. Nenadovic, T. Rajh, and O. I. Micic, *J. Phys. Chem.*, 1985, **89**, 397–399.
3. A. P. Alivisatos, *Science*, 1996, **271**, 933–937.
4. D. Ramadurai, D. Geerpuram, D. Alexson, M. Dutta, N. A. Kotov, Z. Tang, and M. A. Stroscio, *Superlattices Microstruct.*, 2006, **40**, 38–44.
5. L. Brus, *J. Phys. Chem. Solids*, 1998, **59**, 459–465.
6. V. L. Colvin, A. P. Alivisatos, and J. G. Tobin, *Phys. Rev. Lett.*, 1991, **66**, 2786.
7. A. N. Goldstein, C. M. Echer, and A. P. Alivisatos, *Science*, 1992, **256**, 1425–1427.
8. T. Trindade, P. O'Brien, and N. L. Pickett, *Chem. Mater.*, 2001, **13**, 3843–3858.
9. A. Eychmüller, *J. Phys. Chem. B*, 2000, **104**, 6514–6528.
10. R. Rossetti, R. Hull, J. M. Gibson, and L. E. Brus, *J. Chem. Phys.*, 1985, **82**, 552–559.
11. R. Rossetti, J. L. Ellison, J. M. Gibson, and L. E. Brus, *J. Chem. Phys.*, 1984, **80**, 4464–4469.
12. L. E. Brus, *J. Chem. Phys.*, 1984, **80**, 4403–4409.

13. L. Spanhel, M. Haase, H. Weller, and A. Henglein, *J. Am. Chem. Soc.*, 1987, **109**, 5649–5655.
14. T. Vossmeyer, L. Katsikas, M. Giersig, I. G. Popovic, K. Diesner, A. Chemseddine, A. Eychmüller, and H. Weller, *J. Phys. Chem.*, 1994, **98**, 7665–7673.
15. C. B. Murray, D. J. Norris, and M. G. Bawendi, *J. Am. Chem. Soc.*, 1993, **115**, 8706–8715.
16. C. B. Murray, C. R. Kagan, and M. G. Bawendi, *Annu. Rev. Mater. Sci.*, 2000, **30**, 545–610.
17. J. E. Bowen Katari, V. L. Colvin, and A. P. Alivisatos, *J. Phys. Chem.*, 1994, **98**, 4109–4117.
18. I. Mekis, D. V. Talapin, A. Kornowski, M. Haase, and H. Weller, *J. Phys. Chem. B*, 2003, **107**, 7454–7462.
19. D. V. Talapin, A. L. Rogach, A. Kornowski, M. Haase, and H. Weller, *Nano Lett.*, 2001, **1**, 207–211.
20. X. Peng, J. Wickham, and A. P. Alivisatos, *J. Am. Chem. Soc.*, 1998, **120**, 5343–5344.
21. F. Pinaud, D. King, H. P. Moore, and S. Weiss, *J. Am. Chem. Soc.*, 2004, **126**, 6115–6123.
22. L. Qu, and X. Peng, *J. Am. Chem. Soc.*, 2002, **124**, 2049–2055.
23. Y. Ebenstein, T. Mokari, and U. Banin, *Appl. Phys. Lett.*, 2002, **80**, 4033–4035.
24. Z. A. Peng, and X. Peng, *J. Am. Chem. Soc.*, 2002, **124**, 3343–3353.
25. D. V. Talapin, A. L. Rogach, M. Haase, and H. Weller, *J. Phys. Chem. B*, 2001, **105**, 12278–12285.
26. H. Reiss, *J. Chem. Phys.*, 1951, **19**, 482–487.
27. L. Li, H. Qian, N. Fang, and J. Ren, *J. Lumin.*, 2006, **116**, 59–66.
28. D. V. Talapin, A. L. Rogach, E. V. Shevchenko, A. Kornowski, M. Haase, and H. Weller, *J. Am. Chem. Soc.*, 2002, **124**, 5782–5790.
29. M. T. Fernández-Argüelles, W. J. Jin, J. M. Costa-Fernández, R. Pereiro, and A. Sanz-Medel, *Anal. Chimica Acta*, 2005, **549**, 20–25.
30. C.-Y. Chen, C.-T. Cheng, C.-W. Lai, Y.-H. Hu, P.-T. Chou, Y.-H. Chou, and H.-T. Chiu. *Small*, 2005, **1**, 1215–1220.
31. Y. A. Yang, H. Wu, K. R. Williams, and Y. C. Cao., *Angew. Chem. Int. Ed.*, 2005, **44**, 6712–6715.
32. J. J. Li, Y. A. Wang, W. Z. Guo, J. C. Keay, T. D. Mishima, M. B. Johnson, and X. G. Peng, *J. Am. Chem. Soc.*, 2003, **125**, 12567–12575.
33. P. Reiss, J. Bleuse, and A. Pron, *Nano Lett.*, 2002, **2**, 781–784.

34. L. Qu, Z. A. Peng, and X. Peng, *Nano Lett.*, 2001, **1**, 333–337.

35. N. L. Pickett and P. O'Brien, *Chem. Rec.*, 2001, **1**, 467–479.

36. M. A. Malik, N. Revaprasadu, and P. O'Brien, *Chem. Mater.*, 2001, **13**, 913–920.

37. S. L. Cumberland, K. M. Hanif, A. Javier, G. A. Khitrov, G. F. Strouse, S. M. Woessner, and C. S. Yun, *Chem. Mater.*, 2002, **14**, 1576–1584.

38. B. Sun, E. Marx, and N. C. Greenham, *Nano Lett.*, 2003, **3**, 961–963.

39. H. Liu and A. P. Alivisatos, *Nano Lett.*, 2004, **4**, 2397–2401.

40. L. Manna, D. J. Milliron, A. Meisel, E. C. Scher, and A. P. Alivisatos, *Nat. Mater.*, 2003, **2**, 382–385.

41. S. Kumar, and T. Nann, *Small*, 2006, **2**, 316–329.

42. D. V. Talapin, J. H. Nelson, E. V. Shevchenko, S. Aloni, B. Sadtler, and A. P. Alivisatos, *Nano Lett.*, 2007, **7**, 2951–2959.

43. J. A. Gerbec, D. Magana, A. Washington, and G. F. Strouse, *J. Am. Chem. Soc.*, 2005, **127**, 15791–15800.

44. A. L. Washington and G. F. Strouse, *J. Am. Chem. Soc.*, 2008, **130**, 8916–8922.

45. A. L. Washington and G. F. Strouse, *Chem. Mater.*, 2009, **21**, 3586–3592.

46. D. D. Lovingood and G. F. Strouse, *Nano Lett.*, 2008, **8**, 3394–3397.

47. A. V. Firth, Y. Tao, D. S. Wang, J. F. Ding, and F. Bensebaa, *J. Mater. Chem.*, 2005, **15**, 4367–4372.

48. J. Ziegler, A. Merkulov, M. Grabolle, U. Resch-Genger, and T. Nann, *Langmuir*, 2007, **23**, 7751–7759.

49. M. Bruchez, Jr., M. Moronne, P. Gin, S. Weiss, and A. P. Alivisatos, *Science*, 1998, **281**, 2013–2016.

50. D. Gerion, F. Pinaud, S. C. Williams, W. J. Parak, D. Zanchet, S. Weiss, and A. P. Alivisatos, *J. Phys. Chem. B*, 2001, **105**, 8861–8871.

51. W. C. Chan and S. Nie, *Science*, 1998, **281**, 2016–2018.

52. G. P. Mitchell, C. A. Mirkin, and R. L. Letsinger, *J. Am. Chem. Soc.*, 1999, **121**, 8122–8123.

53. B. Dubertret, P. Skourides, D. J. Norris, V. Noireaux, A. H. Brivanlou, and A. Libchaber, *Science*, 2002, **298**, 1759–1762.

54. X. H. Gao, Y. Y. Cui, R. M. Levenson, L. W. K. Chung, and S. M. Nie, *Nat. Biotechnol.*, 2004, **22**, 969–976.

55. N. Tomczak, D. Janczewski, M. Y. Han, and G. J. Vancso, *Prog. Polym. Sci.*, 2009, **34**, 393–430.

56. A. L. Rogach, A. Kornowski, M. Gao, A. Eychmüller, and H. Weller, *J. Phys. Chem. B*, 1999, **103**, 3065–3069.

57. A. E. Raevskaya, A. L. Stroyuk, S. Y. Kuchmiy, Y. M. Azhniuk, V. M. Dzhagan, V. O. Yukhymchuk, and M. Y. Valakh, *Coll. Surf. A: Physicochem. Eng. Asp.*, 2006, **290**, 304–309.

58. H.-Y. Han, Z.-H. Sheng, and J.-G. Liang, *Mater. Lett.*, 2006, **60**, 3782–3785.

59. A. Rogach, S. V. Kershaw, M. Burt, M. Harrison, A. Kornowski, A. Eychmüller, and H. Weller, *Adv. Mater.*, 1999, **11**, 552–555.

60. S. F. Wuister, I. Swart, F. van Driel, S. G. Hickey, and C. de MelloDonega, *Nano Lett.*, 2003, **3**, 503–507.

61. K. Akamatsu, T. Tsuruoka, and H. Nawafune, *J. Am. Chem. Soc.*, 2005, **127**, 1634–1635.

62. S. F. Wuister, C. D. Donega, and A. Meijerink, *J. Phys. Chem. B*, 2004, **108**, 17393–17397.

63. S. F. Wuister, F. van Driel, and A. Meijerink, *Phys. Chem. Chem. Phys.*, 2003, **5**, 1253–1258.

64. A. M. Kapitonov, A. P. Stupak, S. V. Gaponenko, E. P. Petrov, A. L. Rogach, and A. Eychmüller, *J. Phys. Chem. B*, 1999, **103**, 10109–10113.

65. T. Rajh, O. I. Micic, and A. J. Nozik, *J. Phys. Chem.*, 1993, **97**, 11999–12003.

66. L. Li, H. Qian, and J. Ren, *Chem. Comm.*, 2005, 528–530.

67. N. Gaponik, D. V. Talapin, A. L. Rogach, K. Hoppe, E. V. Shevchenko, A. Kornowski, A. Eychmüller, and H. Weller, *J. Phys. Chem. B*, 2002, **106**, 7177–7185.

68. C. L. Li and N. Murase, *Chem. Lett.*, 2005, **34**, 92–93.

69. S. J. Byrne, S. A. Corr, T. Y. Rakovich, Y. K. Gun'ko, Y. P. Rakovich, J. F. Donegan, S. Mitchell, and Y. Volkov, *J. Mat. Chem.*, 2006, **16**, 2896–2902.

70. H. Borchert, D. V. Talapin, N. Gaponik, C. McGinley, S. Adam, A. Lobo, T. Möller, and H. Weller, *J. Phys. Chem. B*, 2003, **107**, 9662–9668.

71. A. Shavel, N. Gaponik, and A. Eychmüller, *J. Phys. Chem. B*, 2006, **110**, 19280–19284.

72. M. Green, H. Harwood, C. Barrowman, P. Rahman, A. Eggeman, F. Festry, P. Dobson, and T. Ng, *J. Mat. Chem.*, 2007, **17**, 1989–1994.

73. H. Bao, E. Wang, and S. Dong, *Small*, 2006, **2**, 476–480.

74. A. L. Rogach, L. Katsikas, A. Kornowski, D. Su, A. Eychmüller, and H. Weller, *Ber. Bunsen-Ges*, 1996, **100**, 1772–1778.

75. A. L. Rogach, T. Franzl, T. A. Klar, J. Feldmann, N. Gaponik, V. Lesnyak, A. Shavel, A. Eychmuller, Y. P. Rakovich, and J. F. Donegan, *J. Phys. Chem. C*, 2007, **111**, 14628–14637.

76. M. T. Harrison, S. V. Kershaw, M. G. Burt, A. Eychmüller, H. Weller, and A. L. Rogach, *Mater. Sci. Eng. B*, 2000, **69–70**, 355–360.
77. Y. Gong, M. Gao, D. Wang, and H. Möhwald, *Chem. Mater.*, 2005, **17**, 2648–2653.
78. H. F. Qian, L. Li, and J. C. Ren, *Mater. Res. Bull.*, 2005, **40**, 1726–1736.
79. Y. He, H. T. Lu, L. M. Sai, Y. Y. Su, M. Hu, C. H. Fan, W. Huang, and L. H. Wang, *Adv. Mater.*, 2008, **20**, 3416.
80. M. P. Moloney, Y. K. Gun'ko, and J. M. Kelly, *Chem. Commun.*, 2007, 3900–3902.
81. M. P. Pileni, *J. Exp. Nanosci.*, 2006, **1**, 13–27.
82. D. Ingert and M. P. Pileni, *Adv. Funct. Mater.*, 2001, **11**, 136–139.
83. N. Pinna, K. Weiss, H. Sack-Kongehl, W. Vogel, J. Urban, and M. P. Pileni, *Langmuir*, 2001, **17**, 7982–7987.
84. M. Z. Rong, M. Q. Zhang, H. C. Liang, and H. M. Zeng, *Appl. Surf. Sci.*, 2004, **228**, 176–190.
85. S. Emin, N. Sogoshi, S. Nakabayashi, M. Villeneuve, and C. Dushkin, *J. Photochem. Photobiol. a-Chem.*, 2009, **207**, 173–180.
86. I. Chakraborty, D. Mitra, and S. P. Moulik, *J. Nanopart. Res.*, 2005, **7**, 227–236.
87. K. A. Gatsouli, S. Pispas, and E. I. Kamitsos, *J. Phys. Chem. C*, 2007, **111**, 15201–15209.
88. C. C. Chen, C. Y. Chao, and Z. H. Lang, *Chem. Mater.*, 2000, **12**, 1516.
89. S. P. Anthony and J. K. Kim, *Chem. Commun.*, 2008, 1193–1195.
90. I. Chakraborty and S. P. Moulik, *J. Nanopart. Res.*, 2004, **6**, 233–240.
91. D. Mitra, I. Chakraborty, and S. P. Moulik, *Colloid J.*, 2005, **67**, 445–450.
92. D. G. Kim, A. Nabeshima, and M. Nakayama, *Jpn. J. Appl. Phys., Part 1-Regular Papers Short Notes & Review Papers*, 2005, **44**, 1514–1517.
93. S. Bhaviripudi, J. Qi, E. L. Hu, and A. M. Belcher, *Nano Lett.*, 2007, **7**, 3512–3517.
94. H. S. Yang, P. H. Holloway, and S. Santra, *J. Chem. Phys.*, 2004, **121**, 7421–7426.
95. Y. Xiang, Y. Y. Zhang, Y. Chang, Y. Q. Chai, J. Wang, and R. Yuan, *Anal. Chem.*, **82**, 1138–1141.
96. M. Rajamathi, and R. Seshadri, *Curr. Opin. Solid State Mater. Sci.*, 2002, **6**, 337–345.
97. Y. D. Li, X. F. Duan, Y. T. Qian, L. Yang, M. R. Ji, and C. W. Li, *J. Am. Chem. Soc.*, 1997, **119**, 7869–7870.

98. Y. T. Qian, *Adv. Mater.*, 1999, **11**, 1101.

99. A. Nag and D. D. Sarma, *J. Nanosci. Nanotechnol.*, 2009, **9**, 5633–5636.

100. Y. D. Li, Y. Ding, Y. T. Qian, Y. Zhang, and L. Yang, *Inorg. Chem.*, 1998, **37**, 2844–2845.

101. Q. Peng, Y. J. Dong, Z. X. Deng, X. M. Sun, and Y. D. Li, *Inorg. Chem.*, 2001, **40**, 3840.

102. S. H. Yu, J. Yang, Z. H. Han, Y. Zhou, R. Y. Yang, Y. T. Qian, and Y. H. Zhang, *J. Mater. Chem.*, 1999, **9**, 1283–1287.

103. U. K. Gautam, M. Rajamathi, F. Meldrum, P. E. D. Morgan, and R. Seshadri, *Chem. Commun.*, 2001, 629–630.

104. Y. Liu, Y. Xu, J. P. Li, B. Zhang, D. Wu, and Y. H. Sun, *Chem. Lett.*, 2004, **33**, 1162–1163.

105. Y. Liu, H. Y. Qiu, Y. Xu, D. Wu, M. J. Li, J. X. Jiang, and G. Q. Lai, *J. Nanopart. Res.*, 2007, **9**, 745–752.

106. U. Pal, P. Santiago, J. Chavez, and J. A. Ascencio, *J. Nanosci. Nanotechnol.*, 2005, **5**, 609–614.

107. H. N. Wang, Z. Y. Guo, and F. L. Du, *Mater. Chem. Phys.*, 2006, **98**, 422–424.

108. G. Ramalingam, N. Melikechi, P. D. Christy, S. Selvakumar, and P. Sagayaraj, *J. Cryst. Growth*, 2009, **311**, 3138–3142.

109. D. F. Underwood, T. Kippeny, and S. J. Rosenthal, *J. Phys. Chem. B*, 2001, **105**, 436–443.

110. M. O'Neil, J. Marohn, and G. McLendon, *J. Phys. Chem.*, 1990, **94**, 4356–4363.

111. D. E. Dunstan, A. Hagfeldt, M. Almgren, H. O. G. Siegbahn, and E. Mukhtar, *J. Phys. Chem.*, 1990, **94**, 6797–6804.

112. S. A. Majetich and A. C. Carter, *J. Phys. Chem.*, 1993, **97**, 8727–8731.

113. A. C. Carter, C. E. Bouldin, K. M. Kemner, M. I. Bell, J. C. Woicik, and S. A. Majetich, *Phys. Rev. B*, 1997, **55**, 13822.

114. M. G. Bawendi, P. J. Carroll, W. L. Wilson, and L. E. Brus, *J. Chem. Phys.*, 1992, **96**, 946–954.

115. H. Fu, and A. Zunger, *Phys. Rev. B*, 1997, **56**, 1496.

116. H. Yang, C. Huang, X. Li, R. Shi, and K. Zhang, *Mater. Chem. Phys.*, 2005, **90**, 155–158.

117. J. Guo, W. L. Yang, and C. C. Wang, *J. Phys. Chem. B*, 2005, **109**, 17467–17473.

118. N. Pradhan, D. Reifsnyder, R. Xie, J. Aldana, and X. Peng, *J. Am. Chem. Soc.*, 2007, **129**, 9500–9509.

119. E. J. Holmyard, and W. G. Palmer, Inorganic Chemistry, Dent and Sons, London, 1952.

120. B. O. Dabbousi, J. Rodriguez-Viejo, F. V. Mikulec, J. R. Heine, H. Mattoussi, R. Ober, K. F. Jensen, and M. G. Bawendi, *J. Phys. Chem. B*, 1997, **101**, 9463–9475.

121. M. A. Malik, P. O'Brien, and N. Revaprasadu, *Chem. Mater.*, 2002, **14**, 2004–2010.

122. M. Nirmal, and L. Brus, *Acc. Chem. Res.*, 1999, **32**, 407–414.

123. H. Zhang, Z. Zhou, B. Yang, and M. Y. Gao, *J. Phys. Chem. B*, 2003, **107**, 8–13.

124. H. B. Bao, Y. J. Gong, Z. Li, and M. Y. Gao, *Chem. Mater.*, 2004, **16**, 3853–3859.

125. J. Aldana, Y. A. Wang, and X. Peng, *J. Am. Chem. Soc.*, 2001, **123**, 8844–8850.

126. M. Jones, J. Nedeljkovic, R. J. Ellingson, A. J. Nozik, and G. Rumbles, *J. Phys. Chem. B*, 2003, **107**, 11346–11352.

127. Y. Wang, Z. Tang, M. A. Correa-Duarte, L. M. Liz-Marzan, and N. A. Kotov, *J. Am. Chem. Soc.*, 2003, **125**, 2830–2831.

128. S. J. Byrne, S. A. Corr, T. Y. Rakovich, Y. K. Gun'ko, Y. P. Rakovich, J. F. Donegan, S. Mitchell, and Y. Volkov, *J. Mater. Chem.*, 2006, **16**, 2896–2902.

129. Y. Wang, Z. Tang, M. A. Correa-Duarte, I. Pastoriza-Santos, M. Giersig, N. A. Kotov, and L. M. Liz-Marzan, *J. Phys. Chem. B*, 2004, **108**, 15461–15469.

130. X. Wang, L. Qu, J. Zhang, X. Peng, and M. Xiao, *Nano Lett.*, 2003, **3**, 1103–1106.

131. S. F. Wuister, and A. Meijerink, *J. Lumin.*, 2003, **105**, 35–43.

132. A. A. Bol, and A. Meijerink, *J. Phys. Chem. B*, 2001, **105**, 10203–10209.

133. H. Zhang, and B. Yang, *Thin Solid Films*, 2002, **418**, 169–174.

134. A. Henglein, *J. Phys. Chem.*, 1982, **86**, 2291–2293.

135. B. I. Ipe, M. Lehnig, and C. M. Niemeyer, *Small*, 2005, **1**, 706–709.

136. U. Resch, H. Weller, and A. Henglein, *Langmuir*, 1989, **5**, 1015–1020.

137. T. Torimoto, S. Y. Murakami, M. Sakuraoka, K. Iwasaki, K. I. Okazaki, T. Shibayama, and B. Ohtani, *J. Phys. Chem. B*, 2006, **110**, 13314–13318.

138. S. R. Cordero, P. J. Carson, R. A. Estabrook, G. F. Strouse, and S. K. Buratto, *J. Phys. Chem. B*, 2000, **104**, 12137–12142.

139. H. Zhang, D. Wang, and H. Möhwald, *Angew. Chem. Int. Ed.*, 2006, **45**, 748–751.

140. Z. Tang, N. A. Kotov, and M. Giersig, *Science*, 2002, **297**, 237–240.

141. Q. Pang, L. J. Zhao, Y. Cai, D. P. Nguyen, N. Regnault, N. Wang, S. H. Yang, W. K. Ge, R. Ferreira, G. Bastard, and J. N. Wang, *Chem. Mater.*, 2005, **17**, 5263–5267.

142. T. Mokari, E. Rothenberg, I. Popov, R. Costi, and U. Banin, *Science*, 2004, **304**, 1787–1790.

143. Y. H. Zhang, H. S. Zhang, M. Ma, X. F. Guo, and H. Wang, *Appl. Surf. Sci.*, 2009, **255**, 4747–4753.

144. C. H. Fischer, and A. Henglein, *J. Phys. Chem.*, 1989, **93**, 5578–5581.

145. M. Gao, S. Kirstein, H. Mohwald, A. Rogach, A. Kornowski, A. Eychmüller, and H. Weller, *J. Phys. Chem. B*, 1998, **102**, 8360–8363.

146. J. Aldana, N. Lavelle, Y. Wang, and X. Peng, in *J. Am. Chem. Soc.*, 2005, 2496–2504.

147. I. G. Dance, M. L. Scudder, and R. Secomb, *Inorg. Chem.*, 1983, **22**, 1794–1797.

148. S. J. Byrne, Y. Williams, A. Davies, S. A. Corr, A. Rakovich, Y. K. Gun'ko, Y. R. Rakovich, J. F. Donegan, and Y. Volkov, *Small*, 2007, **3**, 1152–1156.

149. S. Wang, N. Mamedova, N. A. Kotov, W. Chen, and J. Studer, *Nano Lett.*, 2002, **2**, 817–822.

150. D. Gerion, F. Pinaud, S. C. Williams, W. J. Parak, D. Zanchet, S. Weiss, and A. P. Alivisatos, *J. Phys. Chem. B*, 2001, **105**, 8861–8871.

151. W. J. Parak, D. Gerion, D. Zanchet, A. S. Woerz, T. Pellegrino, C. Micheel, S. C. Williams, M. Seitz, R. E. Bruehl, Z. Bryant, C. Bustamante, C. R. Bertozzi, and A. P. Alivisatos, *Chem. Mater.*, 2002, **14**, 2113–2119.

152. C. Y. Zhang, H. C. Yeh, M. T. Kuroki, and T. H. Wang, *Nat. Mater.*, 2005, **4**, 826–831.

153. M. E. Akerman, W. C. W. Chan, P. Laakkonen, S. N. Bhatia, and E. Ruoslahti, *Proc. Natl. Acad. Sci. U. S. A.*, 2002, **99**, 12617–12621.

154. J. M. Slocik, J. T. Moore, and D. W. Wright, *Nano Lett.*, 2002, **2**, 169–173.

155. H. Mattoussi, J. M. Mauro, E. R. Goldman, G. P. Anderson, V. C. Sundar, F. V. Mikulec, and M. G. Bawendi, *J. Am. Chem. Soc.*, 2000, **122**, 12142–12150.

156. H. Mattoussi, J. M. Mauro, E. R. Goldman, T. M. Green, G. P. Anderson, V. C. Sundar, and M. G. Bawendi, *Phys. Stat. Sol. B*, 2001, **224**, 277–283.

157. E. R. Goldman, E. D. Balighian, H. Mattoussi, M. K. Kuno, J. M. Mauro, P. T. Tran, and G. P. Anderson, *J. Am. Chem. Soc.*, 2002, **124**, 6378–6382.

158. A. M. Derfus, W. C. W. Chan, and S. N. Bhatia, *Adv. Mater.*, 2004, **16**, 961.

159. J. K. Jaiswal, H. Mattoussi, J. M. Mauro, and S. M. Simon, *Nat. Biotech.*, 2003, **21**, 47–51.

160. S. J. Rosenthal, I. Tomlinson, E. M. Adkins, S. Schroeter, S. Adams, L. Swafford, J. McBride, Y. Wang, L. J. DeFelice, and R. D. Blakely, *J. Am. Chem. Soc.*, 2002, **124**, 4586–4594.

161. M. Xie, H. H. Liu, P. Chen, Z. L. Zhang, X. H. Wang, Z. X. Xie, Y. M. Du, B. Q. Pan, and D. W. Pang, *Chem. Comm.*, 2005, 5518–5520.

162. X. Wu, H. Liu, J. Liu, K. N. Haley, J. A. Treadway, J. P. Larson, N. Ge, F. Peale, and M. P. Bruchez, *Nat. Biotech.*, 2003, **21**, 41–46.

163. S. J. Clarke, C. A. Hollmann, Z. Zhang, D. Suffern, S. E. Bradforth, N. M. Dimitrijevic, W. G. Minarik, and J. L. Nadeau, *Nat. Mater.*, 2006, **5**, 409–417.

164. J. O. Winter, T. Y. Liu, B. A. Korgel, and C. E. Schmidt, *Adv. Mater.*, 2001, **13**, 1673–1677.

165. M. Dahan, S. Levi, C. Luccardini, P. Rostaing, B. Riveau, and A. Triller, *Science*, 2003, **302**, 442–445.

166. S. Pathak, S. K. Choi, N. Arnheim, and M. E. Thompson, *J. Am. Chem. Soc.*, 2001, **123**, 4103–4104.

167. P. Sharma, S. Brown, G. Walter, S. Santra, and B. Moudgil, *Adv. Colloid Interface Sci.*, 2006, **123–126**, 471–485.

168. R. Weissleder, *Nat. Biotech.*, 2001, **19**, 316–317.

169. J. M. Klostranec, and W. C. W. Chan, *Adv. Mater.*, 2006, **18**, 1953–1964.

170. A. Becker, C. Hessenius, K. Licha, B. Ebert, U. Sukowski, W. Semmler, B. Wiedenmann, and C. Grötzinger, *Nat. Biotech.*, 2001, **19**, 327–331.

171. R. E. Bailey, and S. Nie, *J. Am. Chem. Soc.*, 2003, **125**, 7100–7106.

172. L. Bakueva, I. Gorelikov, S. Musikhin, X. S. Zhao, E. H. Sargent, and E. Kumacheva, *Adv. Mater.*, 2004, **16**, 926–929.

173. S. Kim, Y. T. Lim, E. G. Soltesz, A. M. De Grand, J. Lee, A. Nakayama, J. A. Parker, T. Mihaljevic, R. G. Laurence, D. M. Dor, L. H. Cohn, M. G. Bawendi, and J. V. Frangioni, *Nat. Biotech.*, 2004, **22**, 93–97.

174. S. W. Kim, J. P. Zimmer, S. Ohnishi, J. B. Tracy, J. V. Frangioni, and M. G. Bawendi, *J. Am. Chem. Soc.*, 2005, **127**, 10526–10532.

175. M. J. Levene, D. A. Dombeck, K. A. Kasischke, R. P. Molloy, and W. W. Webb, *J. Neurophysiol.*, 2004, **91**, 1908–1912.

176. D. R. Larson, W. R. Zipfel, R. M. Williams, S. W. Clark, M. P. Bruchez, F. W. Wise, and W. W. Webb, *Science*, 2003, **300**, 1434–1436.

Chapter 2

Biocompatible CdSe–ZnS Core–Shell Quantum Dots

Ibrahim Yildiz and Françisco M. Raymo*

Department of Chemistry, University of Miami, 1301 Memorial Drive, Coral Gables, Florida 33146-0431, USA

*fraymo@miami.edu

2.1 Semiconductor Quantum Dots

Semiconductor quantum dots are inorganic particles with nanoscale dimensions and outstanding photophysical properties [1–5]. Specifically, these nanoparticles absorb electromagnetic radiations with huge molar extinction coefficients and two-photon absorption cross sections (ε and α in Table 2.1) over broad and continuous ranges of wavelengths [6–12]. Furthermore, they emit electromagnetic radiations upon excitation with luminescence quantum yields (ϕ in Table 2.1) that are comparable to those of conventional organic dyes [6, 13–21]. However, the emission bands of quantum dots are generally symmetric and definitely narrower ($\Delta\lambda$ in Table 2.1) than those of their organic counterparts. In addition, these emission bands can precisely be positioned across the

Quantum Dot Sensors: Technology and Commercial Applications
Edited by John F. Callan and Françisco M. Raymo

ISBN 978-981-4316-00-2 (Hardcover), 978-981-4364-61-4 (eBook)
www.panstanford.com

electromagnetic spectrum, from the ultraviolet to the infrared region, by adjusting the elemental composition and physical dimensions of the nanoparticles. For example, Fig. 2.1 shows the ranges of wavelengths where typical quantum dots emit upon excitation. In each case, the position of the emission band can be regulated within the ranges indicated by acting on the particle diameter. In fact, mixtures of quantum dots, differing in their elemental composition and/or diameter, can be engineered to emit at distinct wavelengths, offering the opportunity to probe multiple parameters in parallel on the basis of luminescence measurements [22]. Moreover, quantum dots have better photobleaching resistances [23] and longer luminescence lifetimes (τ in Table 2.1) than conventional organic dyes [14–16, 18–19, 24–27]. These properties are particularly convenient in the context of bioimaging applications. In fact, the outstanding photobleaching resistances permit the visualization of biological samples for prolonged periods of time [23] and the long luminescence lifetimes offer the opportunity to filter the characteristic short-lived autofluorescence of biological samples with time-gated measurements [28, 29]. Indeed, the unique photophysical signature of semiconductor quantum dots has encouraged their application to a diversity of biological imaging and sensing protocols in alternative to conventional organic fluorophores [30–35].

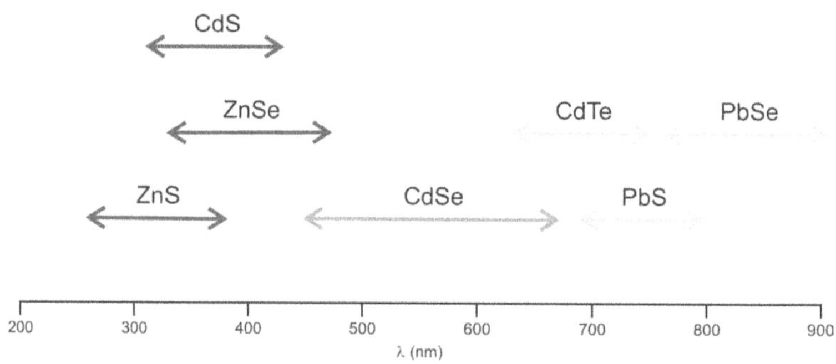

Figure 2.1 Wavelength ranges where typical semiconductor quantum dots emit upon excitation.

2.2 CdSe–ZnS Core–Shell Quantum Dots

The pyrolysis of organometallic reagents in a coordinating solvent can be exploited to prepare nearly monodisperse CdSe particles

Table 2.1 Photophysical parameters of typical organic dyes and quantum dots

	ε[a] ($mM^{-1}\,cm^{-1}$)	α[b] (GM)	λ_{Ex}[c] (nm)	λ_{Em}[d] (nm)	$\Delta\lambda$[e] (nm)	ϕ[f]	τ[g] (ns)
Fluorescein	83	54	492	515	44	0.93	4.0
Oregon Green 488	81	—	491	515	58	0.90	4.1
Cy3	150	—	548	562	50	0.15	0.2
Rhodamine 6G	95	6	550	576	45	0.95	4.1
Texas Red	116	—	583	603	37	0.54	4.2
Cy5	250	400	646	664	40	0.28	2.1
CdS–ZnS	—	—	—	459	24	0.30	—
ZnSe–ZnS	—	—	350	—	420	0.32	—
CdSe–ZnS	98	47,000	541	552	31	0.28	17.1
CdTe	—	—	390	630	55	0.10	—
PbS	—	—	770	920	—	0.50	2,000
PbSe	—	—	990	1400	—	0.81	360

[a] Molar extinction coefficients of the organic dyes (refs. 6, 10) and quantum dots (ref. 8) at λ_{Ex}.

[b] Two-photon absorption cross section of fluorescein at 800 nm (ref. 7), rhodamine 6G at 798 nm (ref. 12), Cy5 at 820 nm (ref. 11) and CdSe–ZnS core–shell quantum dots at 900 nm (ref. 9).

[c] Excitation wavelengths of the organic dyes (refs. 6, 10) and wavelength for the band-gap absorption of the quantum dots (refs. 14–16, 18–19).

[d] Emission wavelengths of the organic dyes (refs. 6, 10) and quantum dots (refs. 14–16, 18–19).

[e] Width at half maximum of the emission bands for the organic dyes (ref. 10) and quantum dots (refs. 14–16, 18–19).

[f] Quantum yields of the organic dyes (refs. 6, 13, 17) and quantum dots (refs. 14–16, 18–19).

[g] Luminescence lifetimes of the organic dyes (refs. 10, 24–27) and average luminescence lifetime of the quantum dots (refs. 14–16, 18–19).

with nanoscale diameters [36, 37]. The resulting quantum dots emit radiations in the visible region of the electromagnetic spectrum (Fig. 2.1), albeit with poor quantum yields. Indeed, defects on the surface of these particles facilitate the nonradiative deactivation of their excited state and are predominantly responsible for the low emission intensities [38]. Nonetheless, the epitaxial growth of a protective ZnS shell around a preformed CdSe core can eliminate defects on the CdSe surface and enhance the luminescence quantum yield significantly [39]. As a result, quantum dots with this particular elemental composition are one of the most studied classes of nanoparticles [40]. In fact, reliable synthetic protocols for the efficient preparation of CdSe–ZnS core–shell quantum dots have already been established. In a typical procedure (Fig. 2.2) [41], CdO and Se are heated at high temperature in a mixture of tetra-*n*-decylphosphonic acid (TDPA), tri-*n*-octylphosphine oxide (TOPO) and tri-*n*-octylphosphine (TOP). This treatment encourages the nucleation of CdSe seeds and their subsequent growth into nanoparticles with TDPA, TOPO and TOP molecules adsorbed on their surface. Once a desired particle size is reached, the temperature can be lowered to arrest growth and the resulting quantum dots can be isolated on the basis of reiterative sequences of precipitation, centrifugation and filtration steps. Then, the assembled CdSe nanoparticles can again be heated at high temperature in a mixture of TOPO and TOP and in the presence of stoichiometric amounts of diethyl zinc and hexamethyldisilathiane [42]. Under these conditions, ZnS layers grow spontaneously around the preformed CdSe particles to generate CdSe–ZnS core–shell quantum dots with TOP and TOPO molecules adsorbed on their surface. After lowering the temperature down to ambient conditions, the final nanostructured assemblies can be isolated, once again, relying on reiterative sequences of precipitation, centrifugation and filtration steps.

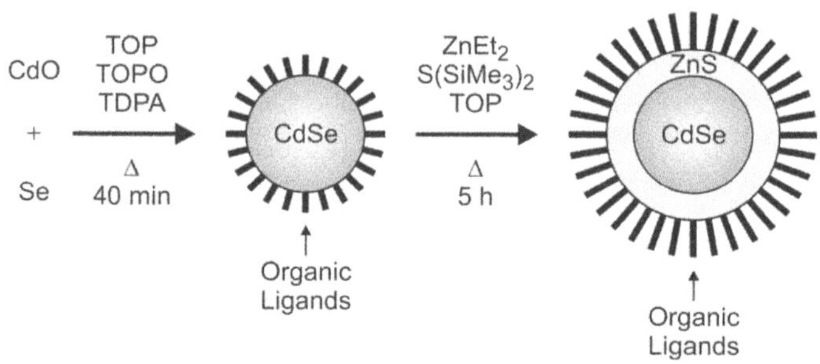

Figure 2.2 Synthesis of CdSe–ZnS core–shell quantum dots coated with the organic ligands tetra-*n*-decylphosphonic acid (TDPA), tri-*n*-octylphosphine oxide (TOPO) and tri-*n*-octylphosphine (TOP).

2.3 Ligand Exchange

Most of the synthetic procedures developed so far for the preparation of CdSe–ZnS core–shell quantum dots rely on the ability of organic surfactants (e.g., TOP and TOPO) to adsorb on the nanoparticle surface and stabilize the resulting assemblies [40]. Because of the hydrophobic organic layer around them, however, these quantum dots are only soluble in nonpolar solvents and cannot be applied to biological preparations. In order to ensure aqueous solubility, the hydrophobic surfactants can be exchanged with hydrophilic ligands. In particular, the affinity of thiol groups for the ZnS shell of the nanoparticles can be exploited to displace the native surfactants and impose hydrophilic character on the resulting nanostructures (Fig. 2.3) [43–49]. In a typical procedure [47], an excess of mercaptoacetic acid is added to a chloroform dispersion of preformed CdSe–ZnS core–shell quantum dots, coated with TDPA, TOPO and TOP. The resulting mixture is then heated under reflux for a sufficient period of time to ensure the displacement of the hydrophobic ligands by mercaptoacetic acid on the nanoparticles surface. After cooling down to ambient temperature, the now hydrophilic quantum dots precipitate and can be isolated after centrifugation and filtration.

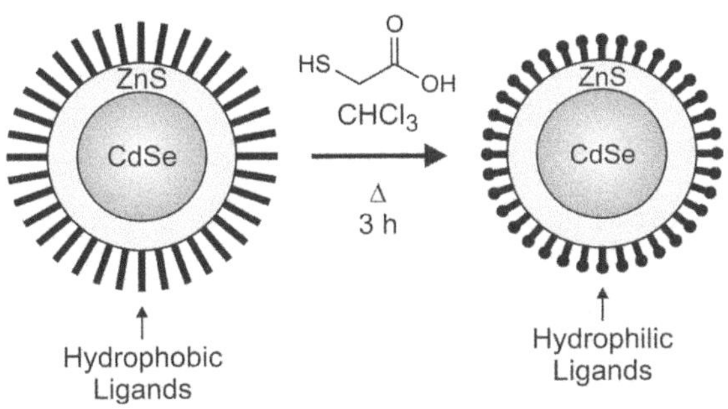

Figure 2.3 Exchange of the native hydrophobic ligands of CdSe–ZnS core–shell quantum dots with hydrophilic thiols.

The adsorption of mercaptocarboxylic acids on the surface of CdSe–ZnS core–shell quantum dots imposes solubility in aqueous environments on the resulting assemblies [43–49]. However, a single thiol group is responsible for the attachment of each carboxylic acid to the nanoparticle surface. As a result, these hydrophilic ligands tend to desorb gradually form the ZnS shell and, therefore, the quantum dots eventually aggregate and precipitate [44]. The transition from monodentate to bidentate ligands prevents desorption and results in the formation of hydrophilic nanoparticles with long-term stabilities [50]. Specifically, dihydrolipoic acid (**1** in Fig. 2.4) is a particularly convenient bidentate and hydrophilic ligand for the preparation of stable and water-soluble quantum dots. Nonetheless, the protic nature of carboxylic acids imposes pH dependence on the solubility of the nanoparticles. At neutral or basic pH, the hydrophilic heads of the ligands are deprotonated and the resulting carboxylate anions ensure solubility. At acidic pH, the carboxylate groups are converted into carboxylic acids and the nanoparticles aggregate and precipitate [49, 51].

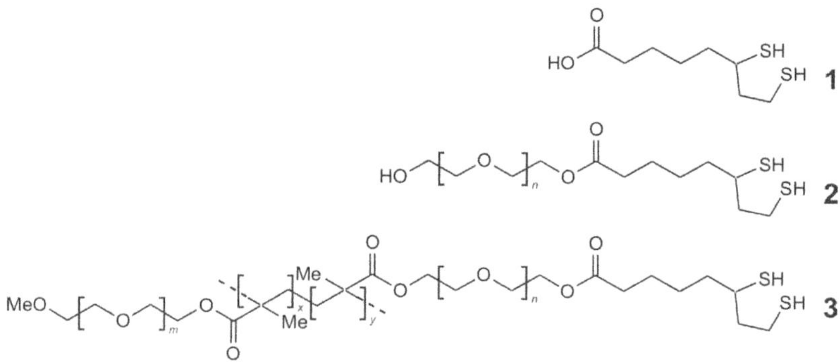

Figure 2.4 Multidentate and hydrophilic ligands for CdSe–ZnS core–shell quantum dots.

The esterification of **1** with polyethylene glycol (**2** in Fig. 2.4) eliminates the pH sensitivity of the ligand, while maintaining the hydrophilic character [51]. Indeed, CdSe–ZnS core–shell quantum dots coated with **1** aggregate at pH lower than 7, while similar nanoparticles passivated with **2** remain dispersed in solution even at a pH of 5. Nonetheless, the transition from organic to aqueous environments has a depressive effect on the luminescence quantum yield of the nanoparticles. Specifically, the quantum yield of CdSe–ZnS core–shell quantum dots, coated with TDPA, TOPO and TOP, is *ca.* 0.8 in toluene and decreases to *ca.* 0.3 after the exchange of the hydrophobic ligands with either **1** or **2** and dispersion of the resulting hydrophilic nanoparticles in water.

Multiple copies of the bidentate fragment of **1** and **2**, as well as multiple polyethylene glycol chains, can be appended to a common macromolecular backbone to generate a polymeric ligand with several anchoring groups and hydrophilic tails (**3** in Fig. 2.4) [52]. The resulting macromolecule can displace the hydrophobic surfactants of preformed CdSe–ZnS core–shell quantum dots and adsorb on the nanoparticle surface. The many thiol groups attached to very same polymer backbone prevent ligand desorption and ensure long-term stability. Similarly, the many polyethylene glycol tails of the polymeric ligand impose hydrophilic character on the nanostructured construct. In fact, quantum dots coated with this particular polymer are stable in aqueous environments for days over a broad pH range (5.0–9.0). Furthermore, the macromolecular backbone, wrapped around the ZnS shell, does not lead to a significant increase in hydrodynamic diameter. Indeed, the hydrodynamic diameter of quantum dots coated with **3** is only 15 nm and is remarkably close to that of nanoparticles coated with **1** (*cf.*, 13 nm). However, the polymeric ligand **3** protects the luminescent core from the aqueous environment and raises the luminescence quantum yield to *ca.* 0.4 (*cf.*, *ca.* 0.3 for **1**).

An alternative approach to encapsulate a CdSe–ZnS core–shell quantum dot in a polymer envelope relies on the adsorption of appropriate monomers on the nanoparticle surface and their subsequent polymerization [53, 54]. In a typical procedure (Fig. 2.5) [53b], preformed quantum dots, coated with TDPA, TOPO and TOP, are dispersed in methanol together with an excess of (3-mercaptopropyl)trimethoxysilane and tetramethylammonium hydroxide and heated for a sufficient period of time to ensure the exchange of the hydrophobic ligands with the mercaptosilane. Then, the resulting nanoparticles are combined with (3-trihydroxysilylpropyl)methyl phosphonate and additional (3-mercaptopropyl)trimethoxysilane to encourage the formation of a polysiloxane envelope with pendant phosphonate groups around each quantum dot. The charged phosphonate groups impose hydrophilic character and water solubility on the final assemblies, while the polysiloxane layer protects the luminescent core form the aqueous environment. In fact, the photophysical response of aqueous dispersions of these quantum dots remains essentially unaffected for days. At least a fivefold decrease in luminescence quantum yield, however, is observed after this treatment and the quantum efficiency of the final assemblies barely approaches 0.2. In addition, the growth of a polysiloxane coating around the inorganic core–shell assembly enlarges considerably the physical dimensions of these constructs, relative to those prepared relying exclusively on ligand exchange. Specifically, atomic force microscopic measurements of identical CdSe–ZnS core–shell quantum dots coated with either a polysiloxane shell or 3-mercaptopropionic acid indicate differences of up to 5 nm in diameter.

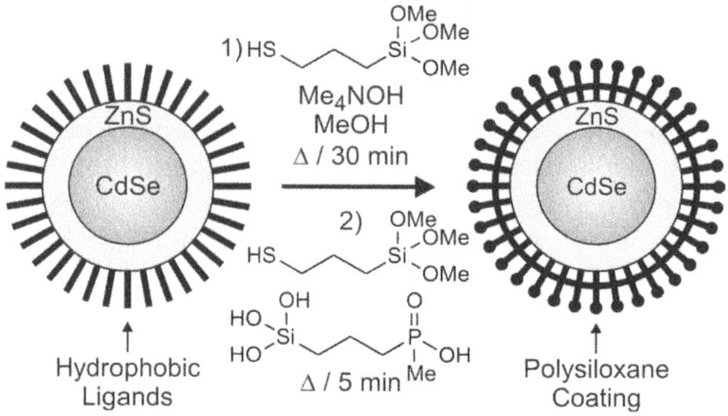

Figure 2.5 Exchange of the native hydrophobic ligands of CdSe–ZnS core–shell quantum dots with a polysiloxane coating.

In alternative to carboxylic acids and polyethylene glycol chains, other hydrophilic groups can be attached to preformed CdSe–ZnS core–shell quantum dots, relying on thiol anchoring groups [48, 55–59]. In particular, the intrinsic biocompatibility of carbohydrates and peptides is especially convenient for the preparation of luminescent nanoparticles for biological appli- cations. For example, an aliphatic tail with a terminal thiol group can be attached to a monosaccharide residue (**4** in Fig. 2.6) and anchored on the surface of CdSe–ZnS core–shell quantum dots [58]. The resulting hydrophilic nanoparticles are readily soluble in aqueous environments, albeit

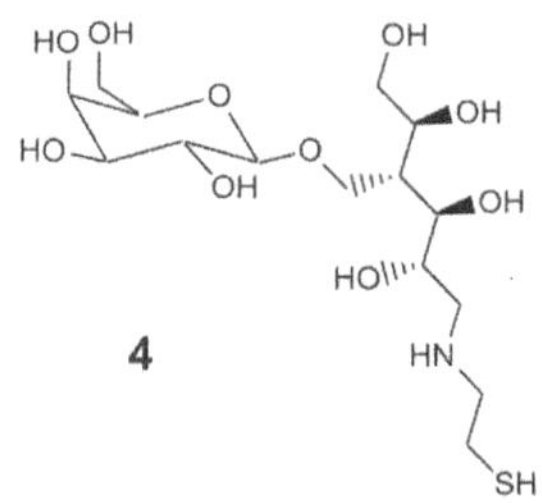

4

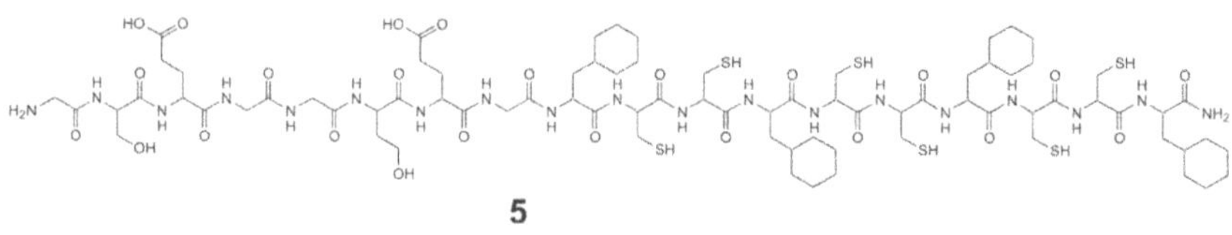

5

Figure 2.6 Monosaccharide- and peptide-based ligands for CdSe–ZnS core–shell quantum dots.

only at neutral pH. Similarly, adhesive and hydrophilic domains can be designed within an oligopeptide sequence (**5** in Fig. 2.6) in order to encourage adsorption on the surface of CdSe–ZnS core–shell quantum dots and ensure water solubility [56]. Indeed, the resulting nanoparticles are soluble in aqueous environments and stable under these conditions over a broad pH range (4.0–10.0). However, transmission electron microscopic and size-exclusion chromatographic measurements indicate that the exchange of the native hydrophobic surfactants of the quantum dots with the hydrophilic peptides leads to an increase in the diameter of the nanostructured constructs of *ca.* 5 nm.

2.4 Ligand Interdigitation

The exchange of the native organic surfactants of preformed CdSe–ZnS core–shell quantum dots with hydrophilic thiols can certainly impose aqueous solubility on the nanoparticles. However, the displacement of the hydrophobic coating has a depressive effect on the luminescence quantum yield. Alternatively, the hydrophobic layer can be left in place to preserve the photophysical properties of the inorganic core and can be overcoated with an additional organic layer of hydrophilic components to ensure water solubility [60]. In a typical procedure [60b], preformed and hydrophobic CdSe–ZnS core–shell quantum dots are combined with a phospholipid–polyethylene glycol conjugate (**6** in Fig. 2.7) in chloroform and maintained at ambient temperature for several hours. Under these conditions, the hydrophobic oligomethylene tails of the phospholipid interdigitate with the oligomethylene tails of the surfactants adsorbed on the nanoparticle surface. These supramolecular interactions result in the noncovalent attachment of polyethylene glycol tails to the quantum dots and culminate with the formation of hydrophilic nanoparticles. The encapsulation of the CdSe–ZnS assembly within the phospholipid envelop has only a minor influence on the luminescence quantum yield. Indeed, nanoparticles with a quantum efficiency of *ca.* 0.4 in water can be prepared, relying on this experimental protocol. However, the hydrodynamic diameter increase from only 7 up 22 nm after the assembly of the phospholipid ligands around the quantum dots. Thus, this approach leads to a significant enhancement in the physical dimensions of the luminescent nanoparticles.

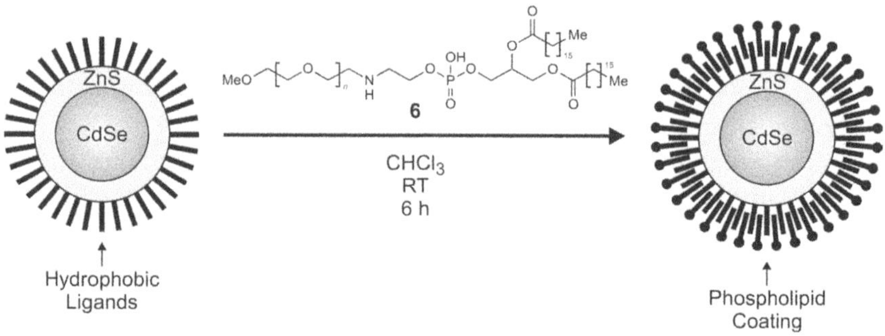

Figure 2.7 Adsorption of hydrophilic phospholipids on the hydrophobic layer of CdSe–ZnS core–shell quantum dots.

In alternative to amphiphilic monomeric ligands, also polymers (e.g., **7–9** in Fig. 2.8) with hydrophilic and hydrophobic domains within their macromolecular backbone can be employed to encapsulate preformed CdSe–ZnS core–shell quantum dots and produce water-soluble luminescent nanoparticles [61–69]. In particular, hydro-philic polyethylene glycol chains and/or carboxylic acids can be mounted on a polymer backbone together with hydrophobic oligomethylene tails. The hydrophobic tails interdigitate with the hydrophobic ligands on the surface of the quantum dots, encouraging the adsorption of the polymers on the nanoparticles. The pendant hydrophilic groups impose aqueous solubility on the resulting assemblies. In addition, the polymeric coating protects the inorganic core form the aqueous environment and preserves its photophysical properties. For example, the absorption and emission spectra, as well as the luminescence quantum yield, of quantum dots coated with **7** do not vary over broad ranges of pH (1–14), salt concentrations (0.01–1 M) and even in the presence of hydrochloric acid (1.0 M) [61]. However, the polymeric coat around the hydrophobic surfactants almost doubles the physical dimensions of the nanostructured constructs relative to the native quantum dots.

2.5 Bioconjugation

The application of quantum dots to biological preparations requires the luminescent nanoparticles to have hydrophilic character and, often, also to be conjugated to appropriate biomolecules [70].

Hydrophilic coatings with pendant carboxylic acids (e.g., Figs. 2.3, 2.4, and 2.8) offer the opportunity to couple quantum dots to proteins on the basis of amide bonds [43, 61, 63, 71–73]. For example, the amphiphilic polymer **9** wraps around preformed CdSe–ZnS core–shell quantum dots, as a result of the interdigitation of its oligomethylene tails with the nanoparticle surfactants. The resulting hydrophilic quantum dots can be dispersed in sodium phosphate buffer at close to neutral pH, together with antibodies raised against the receptor Her-2 of cancer cells and 1-ethyl-3-(3-dimethylami nopropyl)carbodiimide (EDC), and maintained for 2 h at ambient temperature. Under these experimental conditions, the coupling agent EDC encourages the condensation of primary amino groups on the anti-Her2 surface with carboxylic acids on the nanoparticle surface. The final result is the formation of biomolecule–nanoparticle conjugates able to recognize and label cancer cells. Indeed, the incubation of these constructs with human breast cancer cells, overexpressing Her2, resulted in antibody–receptor association and offered the opportunity to visualize the labeled cells, relying on the luminescence of the quantum dots.

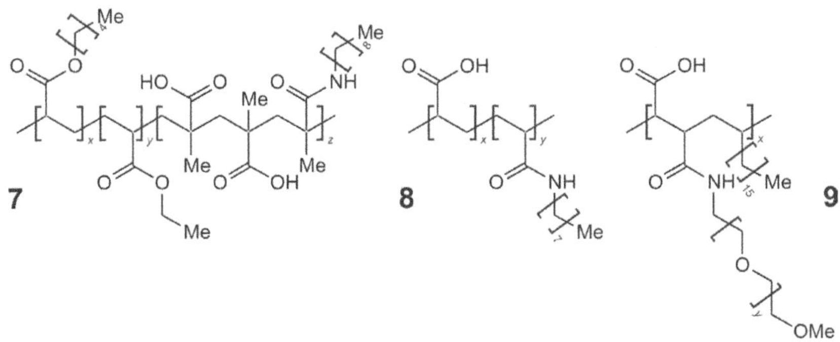

Figure 2.8 Amphiphilic polymeric ligands for CdSe–ZnS core–shell quantum dots.

In alternative to covalent bonding, also electrostatic interactions can be exploited to conjugate biomolecules to quantum dots with carboxylic acids on their surface [73, 74]. At basic pH, the carboxylic acids are converted into carboxylate anions and, hence, the resulting nanoparticles can associated with cationic biomolecules. For example, CdSe–ZnS core–shell quantum dots coated with **1** can be dispersed in sodium borate buffer at a pH of 9. Under these conditions, the carboxylic acids of the ligands adsorbed on the nanoparticle surface are deprotonated to generate negatively charged quantum dots. At this point, the addition of a maltose binding protein homodimer with pendant peptidic chains, incorporating multiple positively charged residues, results in the electrostatic association of the quantum dots with the biomolecules. Interestingly, the nanoparticles retain their photophysical properties after conjugation and their close proximity to the proteins leads to an enhancement in luminescence intensity. Similarly, the biomolecules retain their binding affinity after conjugation and recognize maltose with high association constant.

The affinity of thiol groups for CdSe–ZnS core–shell quantum dots can be exploited to adsorb biomolecules directly on the nanoparticle surface [44]. In particular, preformed quantum dots must first be coated with hydrophilic ligands and then dispersed in aqueous solution in the presence of thiolated biomolecules. For example, bovine serum albumin can be reacted with 2-iminothiolane to introduce an average of 20 thiol groups per protein and then combined with nanoparticles coated with mercaptoacetic acid to generate biomolecule–quantum dot conjugates. Similarly, the adsorption of oligohistidine sequences on CdSe–ZnS core–shell quantum dots can also be exploited to attach biomolecules to the nanoparticle surface [75]. Once again, the quantum dots must first be coated with hydrophilic ligands and then dispersed in aqueous solution in the presence of thiolated biomolecules. For example, an oligonucleotide strand can be extended with a peptide, composed of six histidine residues, and then combined with quantum dots coated with **1**. The final result is the robust attachment of the oligonucleotide–peptide construct to the nanoparticles.

The supramolecular association of biotin and streptavidin together with the affinity of thiol groups for CdSe–ZnS core–shell quantum dots can also be exploited to construct biomolecule–nanoparticle assemblies [73, 76, 77]. In particular, anchoring thiol groups, hydrophilic polyethylene glycol chains and connecting biotin appendages can be integrated within the same molecular skeleton (**10** in Fig. 2.9). The thiol groups ensure the adsorption of the resulting ligands on the nanoparticle surface, while the polyethylene glycol chains impose aqueous solubility. In turn, the biotin terminus can associate with streptavidin linkers conjugated to a protein of interest. For example, the dispersion of CdSe–ZnS core–shell quantum dots, coated with **10**, in aqueous solution, containing streptavidin and biotinylated phycoerythrin, results in the assembly of nanoparticle–streptavidin–phycoerythrin constructs. The further adsorption of cell-penetrating peptides, carrying oligohistidine residues, on the quantum dots offers the opportunity to transport intracellularly the conjugated protein. In fact, the incubation of these nanostructured assemblies with COS-1 cells results in the cellular uptake of phycoerythrin.

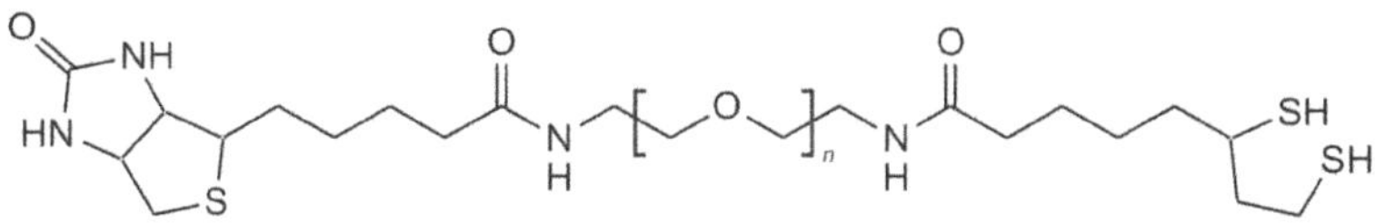

10

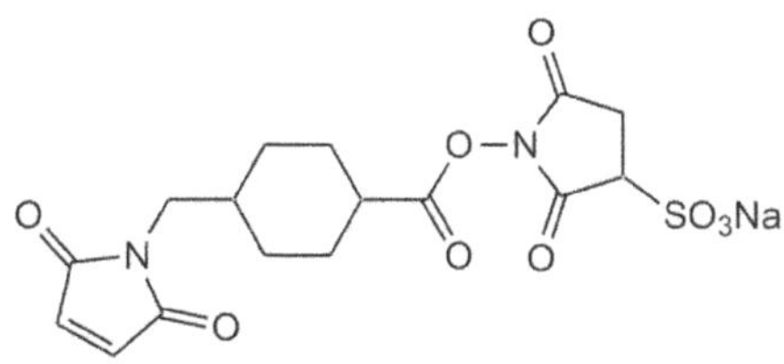

11

Figure 2.9 Biotinylated ligand and bifunctional linkers for the conjugation of CdSe–ZnS core–shell quantum dots to biomolecules.

In these nanoparticle–streptavidin–protein constructs, streptavidin is essentially a bifunctional linker engineered to hold supramolecularly the ternary assembly together. Similarly, relatively small synthetic linkers can be designed to permit the covalent connection of quantum dots to biomolecules. In particular, polysiloxane shells, bearing pendant phosphonate and thiol groups, can be build around preformed CdSe–ZnS core–shell quantum dots, adapting the protocol highlighted in Fig. 2.5 [78]. The charged phosphonate groups ensure water solubility, while the thiol groups can be connected covalently to the maleimido terminus of the bifunctional linker **11** on the basis of a thioether linkage. The other end of **11** has a *N*-hydroxysuccinimide ester, which can react with a primary amino group on the surface of a biomolecule to form an amide bond. Relying on this procedure, single- and double-stranded oligonucleotides can be connected to quantum dots with essentially no influence on the photophysical properties on the inner CdSe core.

2.6 Cytotoxicity

The established toxicity of cadmium [79–81] is a serious concern in view of the possible application of CdSe–ZnS core–shell quantum dots to living organisms. Indeed, the leakage of Cd(II) cations from the core of these luminescent nanoparticles can have deleterious consequences on cells [82–89]. The presence of a ZnS shell around the CdSe core, however, limits cadmium release. According to model poisoning studies of NRK cells, the transition from CdSe, coated with 3-mercaptopropionic acid, to CdSe–ZnS quantum dots, passivated with the same hydrophilic thiol, leads to a tenfold increase in the cytotoxicity concentration threshold [84]. The embedding of the CdSe–ZnS core–shell assembly in a polysiloxane envelope (e.g., Fig. 2.5) dramatically reduces Cd(II) leakage and increases the cytotoxicity concentration threshold even further by, at least, one more order of magnitude. Similarly, organic polymers wrapped around preformed CdSe–ZnS core–shell quantum dots limit Cd(II) release and the resulting nanoparticles have negligible cytotoxicity. For example, the proliferation rate of human umbilical vein endothelial cells (HUVEC), incubated with CdSe–ZnS core–shell quantum dots coated with the hydrophilic polythiol **3**, is essentially identical to that of HUVEC cultured in the absence of quantum dots [52]. Analogously, the incubation of human breast cancer cells SK-BR-3 in the presence of CdSe–ZnS core–shell quantum dots, coated with the amphiphilic polymer **7**, for up to one day resulted in the survival of more than 90% of the SK-BR-3 cells [63]. Thus, the encapsulation of these luminescent nanoparticles with appropriate hydrophilic polymer coatings can overcome the complications associated with the intrinsic toxicity of the core components and ensure biocompatibility for the resulting assemblies.

2.7 Conclusions

Semiconductor quantum dots are promising materials for the realization of luminescent probes with unprecedented performance. Indeed, the outstanding photophysical properties of these inorganic nanoparticles outperform those of organic dyes and it is becoming apparent that quantum dots can replace conventional fluorophores in a diversity of biomedical applications. In particular, nanoparticles with a CdSe core, surrounded by a ZnS shell, appear to be promising luminescent labels for bioimaging studies. Their emission bands can be tuned across the visible region and multicolor assays can be designed around the photophysical signature of these CdSe–ZnS core–shell assemblies. Furthermore, reliable protocols for their efficient preparation are now available and these particular nanoparticles can even be purchased from commercial sources. However, their synthesis relies on the coordinating ability of organic surfactants and the resulting quantum dots are hydrophobic and insoluble in water. The hydrophobic organic coating can be exchanged with hydrophilic thiols to impose aqueous solubility. In particular, mercaptocarboxylic acids can easily be attached to the surface of the ZnS shell to produce water-soluble nanoparticles. Nonetheless, these hydrophilic ligands tend to desorb gradually form the nanoparticle surface leading to aggregation and precipitation of the quantum dots. This problem can be overcome by wrapping polymers with multiple thiol groups along the same macromolecular backbone around the CdSe–ZnS core–shell assembly. However, this treatment has a depressive effect on the luminescence quantum yield. Alternatively, the hydrophobic ligands can be left in place on the ZnS shell and the entire nanostructured construct can be entrapped within an amphiphilic capsule. The resulting quantum dots retain their luminescence intensity, but the physical dimensions of the overall assembly increase considerably. In any case, both strategies generate remarkably stable and water-soluble luminescent nanoparticles and permit their conjugation to a diversity of biomolecules. Furthermore, the encapsulation of the inorganic component inside an organic coating prevents the leakage of toxic Cd(II) cations and, in fact, these particular quantum dots have essentially no cytotoxicity. Thus, the careful design of passivation strategies to coat the surface of preformed CdSe–ZnS core–shell quantum dots with appropriate hydrophilic envelopes is crucial to ensure biocompatibility and permit the application of these promising nanostructured materials to biology.

Acknowledgments

We thank the National Science Foundation (CAREER Award CHE-0237578, CHE-0749840 and CHE-1049860) and the University of Miami for financial support.

References

1. Bawendi, M. G., Steigerwald, M. L., and Brus, L. E. *Ann. Rev. Phys. Chem.*, 1990, **41**, 477–496.
2. Alivisatos, A. P. *Science,* 1996, **271**, 933–937.
3. Yoffe, A. D. *Adv. Physics*, 2001, **50**, 1–208.
4. Efros, Al. L., and Rosen, M. *Ann. Rev. Mater. Sci.*, 2000, **30**, 475–521.
5. Burda, C., Chen, X. B., Narayana, R., and El-Sayed, M. A. *Chem. Rev.*, 2005, **105**, 1025–1102.
6. Mujumdar, R. B., Ernst, L. A., Mujumdar, S. R., Lewis, C. J., and Waggoner, A. S. *Bioconj. Chem.*, 1993, **4**, 105–111.
7. Song, J. M., Inoue, T., Kawazumi, H., and Ogawa T. *Anal. Sci.*, 1999, **15**, 601–603.
8. Yu, W., Qu, L., Guo, W., and Peng, X. *Chem. Mater.*, 2003, **15**, 2854–2860.
9. Larson, D. R., Zipfel, W. R., Williams, R. M., Clark, S. W., Bruchez, M. P., Wise, F. W., and Webb, W. W. *Science,* 2003, **300**, 1434–1436.
10. Haugland, R. P. *The Handbook: A Guide to Fluorescent Probes and Labeling Technologies*; Molecular Probes: Eugene, 2005.
11. Lukomska, J., Gryczynski, I., Malicka, J., Makowiec, S., Lakowicz, J. R., and Gryczynski, Z. J. *Biochem. Biophys. Res. Commun.*, 2005, **328**, 78–84.
12. Kauert, M., Stoller, P. C., Frenz, M., and Rička, J. *Opt. Express.*, 2006, **14**, 8434–8447.
13. Kubin, R. F. and Fletcher, A. N. *J. Luminescence*, 1982, **27**, 455–462.
14. Du, H., Chen, C., Krishnan, R., Krauss, T. D., Harbold, J. M., Wise, F. W., and Thomas, M. G. *Nano Lett.*, 2002, **2**, 1321–1324.
15. Steckel, J. S., Zimmer, J. P., Sullivan, S. C., Stott, N. E., Bulovic, V., and Bawendi, M. G. *Angew. Chem. Int. Ed.*, 2004, **43**, 2154–2158.
16. Chen, H. S., Lo, B., Hwang, J. Y., Chang, G. Y., Chen, C. M., Tasi, S. J., and Jassy, W. *J. Phys. Chem., B* 2004, **108**, 17119–17123.
17. Unruh, J. R., Gokulrangan, G., Wilson, G. S., and Johnson, C. K. *Photochem. Photobiol.*, 2005, **81**, 682–690.
18. Wolcott, A., Gerion, D., Visconte, M., Sun, J., Schwartzberg, A., Chen, S., and Zhang, J. Z. *J. Phys. Chem. B,* 2006, **110**, 5779–5789.
19. Tomasulo, M., Yildiz, I., Kaanumalle, S. L., and Raymo, F. M. *Langmuir* 2006, **24**, 10284–10290.
20. Lakowicz, J. R. *Principles of Fluorescence Spectroscopy*; Springer: New York, 2006.

21. Clark, S. W., Harbold, J. M., and Wise, F. W. *J. Phys. Chem. C,* 2007, **111**, 7302–7305.

22. (a) Kobayashi, H., Koyama, Y., Barrett, T., Hama, Y., Regino, C. A. S., Shin, I. S., Jang, B. S., Le, N., Paik, C. H., Choyke, P. L., and Urano, Y. *ACS Nano*, 2007, **1**, 258–264. (b) Kobayashi, H., Hama, Y., Koyama, Y., Barrett, T., Regino, C. A. S., Urano, Y., and Choyke, P. L. *Nano Lett.*, 2007, **7**, 1711–1716.

23. Howarth, M., Takao, K., Hayashi, Y., and Ting, A. Y. *Proc. Natl. Acad. Sci. USA*, 2005, **102**, 7583–7588.

24. Brismar, H., Trepte, O., and Ulfhake, B. *Histochem. Cytochem.*, 1995, **43**, 699–707.

25. Hanley, Q. S., Subramaniam, V., and Arndt-Jovin, D. J. *Cytometry,* 2001, **43**, 248–260.

26. Tinnefeld, P., Herten, D. P., and Sauer, M. *J. Phys. Chem. A*, 2001, **105**, 7989–8003.

27. Sabanayagam, C. R., Eid, J. S., and Meller, A. *J. Chem. Phys.*, 2005, **122**, 061103-1–061103-5.

28. Dahan, M., Laurence, T., Pinaud, F., Chemla, D. S., Alivisatos, A. P., Sauer, M., and Weiss, S. *Opt. Lett.*, 2001, **26**, 825–827.

29. Grecco, H. E., Lidke, K. A., Heintzmann, R., Lidke, D. S., Spagnuolo, C., Martinez, O. E., Jares-Erijman, E. A., and Jovin, T. M. *Microsci. Res. Tech.*, 2005, **65**, 169–179.

30. Alivisatos, A. P. *Nat. Biotechol.*, 2004, **22**, 47–52.

31. Gao, X., Yang, L., Petros, J. A., Marshall, F. F., Simons, J. W., and Nie, S. *Curr. Op. Biotechnol.*, 2005, **16**, 63–72.

32. Rosi, N. L. and Mirkin, C. A. *Chem. Rev.*, 2005, **105**, 1547–1562.

33. Medintz, I. G., Uyeda, H. T., Goldam, E. R., and Mattoussi, H. *Nat. Mater.*, 2005, **4**, 435–446.

34. Michalet, X., Pinaud, F. F., Bentolila, L. A., Tsay, J. M., Doose, S., Li, J. J., Sundaresan, G., Wu, A. M., Gambhir, S. S., and Weiss, S. *Science*, 2005, **307**, 538–544.

35. Resch-Genger, U., Grabolle, M., Cavaliere-Jaricot, S., Nitschke, R., and Nann, T. *Nat. Methods*, 2008, **5**, 763–775.

36. Murray, C. B., Norris, D. J., and Bawendi, M. G. *J. Am. Chem. Soc.*, 1993, **115**, 8706–8715.

37. Katai, J. E. B., Colvin, V. L., and Alivisatos, A. P. *J. Phys. Chem.*, 1994, **98**, 4109–4117.

38. Underwood, D. F., Kippeny, T. C., and Rosenthal, S. J. *Eur. Phys. J. D* 2001, **16**, 241–244.

39. Hines, M. A. and Guyot-Sionnest, P. *J. Phys. Chem.*, 1996, **100**, 468–471.
40. Rosenthal, S. J., McBride, J., Pennycook, S. J., and Feldman, L. C. *Surf. Sci. Rep.*, 2007, **62**, 111–157.
41. Peng, Z. A. and Peng, X. *J. Am. Chem. Soc.*, 2001, **123**, 183–184.
42. Dabbousi, B. O., Rodriguez-Viejo, J., Mikulec, F. V., Heine, J. R., Mattoussi, H., Ober, R., Jensen, K. F., and Bawendi, M. G. *J. Phys. Chem. B*, 1997, **101**, 9463–9475.
43. Chan, W. C. W. and Nie, S. *Science*, 1998, **281**, 2016–2018.
44. Willard, D. M., Carillo, L. L., Jung, J., and Van Orden, A. *Nano Lett.*, 2001, **1**, 469–474.
45. Kloepfer, J. A., Bradforth, S. E., and Nadeau, J. L. *J. Phys. Chem. B*, 2005, **109**, 9996–10003.
46. Gurinovich, L. I., Artem'ev, M. V., Lyutich, A. A., and Gaponenko, S. V. *J. Appl. Spectrosc.*, 2006, **73**, 572–575.
47. Yildiz, I., Tomasulo, M., and Raymo, F. M. *Proc. Natl. Acad. Sci. USA*, 2006, **103**, 11457–11460.
48. Murcia, M. J., Shaw, D. L., Long, E. C., and Naumann, C. A. *Opt. Commun.*, 2008, **281**, 1771–1780.
49. Algar, W. R. and Krull, U. J. *Chem. Phys. Chem.*, 2007, **8**, 561–568.
50. Mattoussi, H., Mauro, J. M., Goldman, E. R., Anderson, G. P., Sundar, V. C., Mikulec, F. V., and Bawendi, M. G. *J. Am. Chem. Soc.*, 2000, **122**, 12142–12150.
51. Uyeda, H. T., Medintz, I. L., Jaiswal, J. K., Simon, S. M., and Mattoussi, H. *J. Am. Chem. Soc.*, 2005, **127**, 3870–3878.
52. Yildiz, I., McCaughan, B., Cruickshank, S. F., Callan, J. F., and Raymo, F. M. *Langmuir*, 2009, **25**, 7090–7096.
53. (a) Bruchez, M. P., Morrone, M., Gin, P., Weiss, S., and Alivisatos, A. P. *Science*, 1998, **281**, 2013–2016. (b) Gerion, D., Pinaud, F., Williams, S. C., Parak, W. J., Zanchet, D., Weiss, S., and Alivisatos, A. P. *J. Phys. Chem. B*, 2001, **105**, 8861–8871.
54. Jiang, W., Mardyani, S., Fischer, H., and Chan, W. C. W. *Chem. Mater.*, 2006, **18**, 872–878.
55. Sukhanova, A., Devy, J., Venteo, L., Kaplan, H., Artemyev, M., Oleinikov, V., Klinov, D., Pluot, M., Cohen, J. H. M., and Nabiev, I. *Anal. Biochem.*, 2004, **324**, 60–67.
56. Pinaud, F., King, D., Moore, H.-P., and Weiss, S. *J. Am. Chem. Soc.*, 2004, **126**, 6115–6123.
57. Wisher, A. C., Bronstein, I., and Chechik, V. *Chem. Commun.*, 2006, 1637–1639.

58. Babu, P., Sinha, S., and Surolia, A. *Bioconjug. Chem.*, 2007, **18**, 146–151.

59. Callan, J. F., Mulrooney, R. C., and Kamila, S. *J. Fluoresc.*, 2008, **18**, 1157–1161.

60. (a) Dubertret, B., Skourides, P., Norris, D. J., Noireaux, V., Brivanlou, A. H., and Libacher, A. *Science*, 2002, **298**, 1759–1762. (b) Carion, O., Mahler, B., Pons, T., and Dubertret, B. *Nat. Protocols*, 2007, **2**, 2383–2390.

61. Gao, X., Cui, Y., Levenson, R. M., Chung, L. W. K., and Nie, S. *Nat. Biotechnol.*, 2004, **22**, 969–976.

62. Zhou, M., Nakatani, E., Gronenberg, L. S., Tokimoto, T., Wirth, M. J., Hruby, V. J., Roberts, A., Lynch, R. M., and Ghosh, I. *Bioconj. Chem.*, 2007, **18**, 232–332.

63. Nann, T. *Chem. Commun.*, 2005, 1735–1736.

64. (a) Yu, W. W., Chang, E., Drezek, R., and Colvin, V. L. *J. Biomed. Nanotechol.*, 2006, **2**, 225–228. (b) Yu, W. W., Chang, E., Falkner, J. C., Zhang, J., Al-Somali, A. M., Sayes, C. M., Johns, J., Drezek, R., and Colvin, V. L. *J. Am. Chem. Soc.*, 2007, **129**, 2871–2879.

65. (a) Fernández-Argüelles, M. T., Yakovlev, A., Sperling, R. A., Luccardini, C., Gaillard, S., Sanz Medel, A., Mallet, J.-M., Brochon, J.-C., Feltz, A., Oheim, M., and Parak, W. J. *Nano Lett.*, 2007, **7**, 2613–2617. (b) Lin, C.-A. J., Sperling, R. A., Li, J. K., Yang, T.-Y., Li, P.-Y., Zanella, M., Chang, W. H., and Parak, W. J. *Small*, 2008, **4**, 334–341.

66. Anderson, R. E., and Chan, W. C. W. *ACS Nano*, 2008, **2**, 1341–1352.

67. Kairdolf, B. A., Smith, A. M., and Nie, S. *J. Am. Chem. Soc.*, 2008, **130**, 12866–12867.

68. Travert-Branger, N., Dubois, F., Carion, C., Carrot, G., Mahler, B., Dubertret, B., Doris, E., and Mioskowski, C. *Langmuir*, 2008, **24**, 3016–3019.

69. Wua, Z., Zhaoa, Y., Qiub, F., Li, Y., Wanga, S., Yanga, B., Chenc, L., Sunc, J., and Wang, J. *Coll. Surf. A*, 2009, **350**, 121–129.

70. Frasco, M. F. and Chaniotakis, N. *Anal. Bioanal. Chem.*, 2010, **396**, 229–240.

71. So, M.-K., Xu, C., Loening, A. M., Gambhir, S. S., and Rao, J. *Nat. Biotechnol.*, 2006, **24**, 339–343.

72. Liu, W., Howarth, M., Greytak, A. B., Zheng, Y., Nocera, D. G., Ting, A. Y., and Bawendi, M. G. *J. Am. Chem. Soc.*, 2008, **130**, 1274–1284.

73. Xing, Y., Chaudry, Q., Shen, C., Kong, K. Y., Zhau, H. E., Chung, L. W., Petros, J. A., O'Regan, R. M., Yezhelyev, M. V., Simons, J. W., Wang, M. D., and Nie, S. *Nat. Protocol*, 2007, **2**, 1152–1165.

74. (a) Mattoussi, H., Mauro, J. M., Goldman, E. R., Anderson, G. P., Sundar, V. C., Mikulec, F. V., and Bawendi, M. G. *J. Am. Chem. Soc.*, 2000, **122**, 12142–12150. (b) Goldman, E. R., Anderson, G. P., Tran, P. T., Mattoussi, H., Charles, P. T., and Mauro, J. M. *Anal. Chem.*, 2002, **74**, 841–847.

75. (a) Goldman, E. R., Medintz, I. L., Hayhurst, A., Anderson, G. P., Mauro, J. M., Iverson, B. L., Georgiou, G., and Mattoussi, H. *Anal. Chim Acta*, 2005, **534**, 63–67. (b) Medintz, I. L., Berti, L., Pons, T., Grimes, A. F., English, D. S., Alessandrini, A., Facci, P., and Mattoussi, H. *Nano Lett.*, 2007, **7**, 1741–1748.

76. Chen, F. and Gerion, D. *Nano Lett.*, 2004, **4**, 1827–1832.

77. Susumu, K., Uyeda, H. T., Medintz, I. L., Pons, T., Delehanty, J. B., and Mattoussi, H. *J. Am. Chem. Soc.*, 2007, **129**, 13987–13996.

78. Parak, W. J., Gerion, D., Zanchet, D., Woerz, A. S., Pellegrino, T., Micheel, C., Williams, S. C., Seitz, M., Bruehl, R. E., Bryant, Z., Bustamante, C., Bertozzi, C. R., and Alivisatos, A. P. *Chem. Mater.*, 2002, **14**, 2113–2119.

79. Santone, K. S., Acosta, D., and Bruckner, J. V. *J. Toxicol. Environ. Health* 1982, **10**, 169–177.

80. Limaye, D. A. and Shaikh, Z. A. *Toxicol. Appl. Pharmacol.*, 1999, **154**, 59–66.

81. Rikans, L. E. and Yamano, T. *J. Biochem. Mol. Toxicol.*, 2000, **14**, 110–117.

82. Derfus, A. M., Chan, W. C. W., and Bhatia, S. N. *Nano Lett.*, 2004, **4**, 11–18.

83. Hoshino, A., Fujioka, K., Oku, T., Suga, M., Sasaki, Y. F., Ohta, T., Yasuhara, M., Suzuki, K., and Yamamoto, K. *Nano Lett.*, 2004, **4**, 2163–2169.

84. Kirchner, C., Liedl, T., Kudera, S., Pellegrino, T., Javier A. M., Gaub, H. E., Stoelzle, S., Fertig, N., and Parak, W. J. *Nano Lett.*, 2005, **5**, 331–338.

85. Hardman, R. *Environ. Health Persp.*, 2006, **114**, 165–172.

86. (a) Chan, W., and Shiao, N. *Acta Pharmacol. Sin.*, 2008, **28**, 259–266. (b) Hsieh, M.-S., Chan, W., and Shiao, N. *Int. J. Mol. Sci.*, 2009, **10**, 2122–2135.

87. Wang, L., Nagesha, D. K., Selvarasah, S., Dokmeci, M. R., and Carrier, R. L. *J. Nanobiotechnol.*, 2008, **6**, 11-1–15.

88. Karabanovas, V., Zakarevicius, E., Sukackaite, A., Streckyte, G., and Rotomskis, R. *Photochem. Photobiol Sci.*, 2008, **7**, 725–729.

89. Su, Y., He, Y., Lu, H., Sai, L., Li, Q., Li, W., Wang, L., Shen, P., Huang, Q., and Fan, C. *Biomaterials*, 2009, **30**, 19–25.

Chapter 3

Electrochemical Properties of Semiconductor Quantum Dots

Matteo Amelia, Alberto Credi, and Serena Silvi
Dipartimento di Chimica "G. Ciamician",
and Interuniversity Center for the Chemical Conversion of Solar Energy (SolarChem),
Università di Bologna, via Selmi 2, 40126 Bologna, Italy
matteo.amelia@unibo.it, alberto.credi@unibo.it, serena.silvi@unibo.it

3.1 Introduction

The extraordinary spectroscopic properties of semiconductor nanocrystals (quantum dots, QDs) [1–6] make them a promising alternative to organic molecular dyes in a variety of applications [7, 8], spanning analytical sciences, diagnostics, materials science, and energy conversion. By exploiting both self-assembly and ligand exchange methodologies [6, 8, 9], the surface of QDs can be modified with molecular species, enabling the construction of nanocrystal–molecule conjugates [10]. These modular nanoassemblies, in analogy with (supra)molecular devices [11, 12], can exhibit valuable functionalities, resulting from a synergistic integration of the properties of the molecular and nanocrystal components.

Quantum Dot Sensors: Technology and Commercial Applications
Edited by John F. Callan and Françisco M. Raymo

ISBN 978-981-4316-00-2 (Hardcover), 978-981-4364-61-4 (eBook)
www.panstanford.com

In many instances, such functionalities are related to the exchange of electrons, either within the nanoassembly or involving external species [13]. As in the case of molecular devices, the rational design of functional assemblies that exploit electron-transfer processes for their operation requires a prior knowledge of the redox properties of the components; in fact, the fine-tuning of these parameters is essential to achieve the desired function [11, 12, 14–16]. In such a context, the amount of information available on the structural and electronic properties of organic and inorganic molecular components is invaluable. Unfortunately, the same level of understanding has not yet been achieved in the case of semi-conductor nanocrystals. In particular, it is still not entirely clear how the dimension of the core, the thickness of the inorganic shell and the nature of the organic ligands can influence the redox properties of semiconductor nanocrystals. Since their discovery in the 80s, quantum dots have attracted much attention and a large amount of literature is available on their synthesis, properties and applications [6, 17–20]. On the other hand, systematic investigations on the relationship between the structural parameters and the redox properties of QDs are rare [21].

The aim of this chapter is to collect and discuss the results of electrochemical studies carried out on binary semiconductor nanocrystals (core and core–shell) of spherical shape. The chapter starts with a brief overview of the peculiar electronic properties of QDs (Section 3.2), and of the electrochemical techniques commonly used for the study of such systems (Section 3.3). Sections 3.4 and 3.5 contain a description of examples taken from the literature. We will focus on CdSe and CdTe nanocrystals, because the relatively large number of papers dealing with these QDs enable a more insightful discussion on their redox properties. Whenever possible, we will compare the data obtained with different techniques by different groups in order to attempt an interpretation of sometimes contradictory results. Sections 3.6 and 3.7 are dedicated to emerging or potential applications that exploit the electroactivity of quantum dots. Finally, limitations and perspectives of these systems are summarized in Section 3.8.

3.2 Basic Electronic Properties of QDs

Several physical properties of bulk materials change substantially when they are in the form of particles with a size on the order of nanometers [5]. Examples of this behavior are melting points and other parameters related to phase transformations, charging energies, spectroscopic properties, etc. For semiconductors nanoparticles the most noticeable consequence of the size effect is that the band gap — namely, the energy difference between the top edge of the valence band and the bottom edge of the conduction band — can be modified on varying the particle size.

In a semiconductor, absorption of a photon with energy equal or greater than the bandgap results in excitation of an electron from the valence to the conduction band, leaving a hole in the valence band. Such an electron–hole pair, or exciton, is bound by the electrostatic attraction between the opposite charges. The Bohr radius measures the extension of the exciton wavefunction over the crystal lattice; for example, the Bohr radius for excitons in CdSe is about 5 nm. When the size of the particle approaches the Bohr radius, the optical and electrical properties of the material become dependent on its physical dimension, owing to the so-called quantum confinement effect [1–6, 22]. In these conditions, the band structure of the semiconductor changes into discrete levels, and the energy difference between the highest occupied level and the lowest unoccupied level widens as the particle size decreases, following in many cases the behavior expected for an electron inside a three-dimensional box. In some cases a description in terms of molecular orbitals is more appropriate than that of band theory, reflecting the nature of QDs as systems lying between bulk materials and molecular species.

Electrochemical measurements are usually employed to estimate the energy of HOMO and LUMO levels of electroactive molecular species. They can also provide information on the electron-transfer kinetics and give access to other parameters such as diffusion coefficients, interfacial properties, thermodynamic and kinetic constants of reactions coupled with the redox process [15, 16, 23]. Similarly, voltammetric experiments can give information on the absolute energies of the valence and conduction bands of quantum dots. The process of charge transfer in quantum dots can be schematized in three ways (Fig. 3.1) [24]: (a) electron or hole addition [25] to a neutral nanoparticle; (b) simultaneous injection of a hole and an electron in two non-interacting quantum dots; (c) formation of an electron hole pair within the same particle by optical excitation. The charging energy for addition of one electron (Fig. 3.1a) is given by Eq. 3.1:

$$\mu_1 \equiv E_1[e_1] - E_0 = \varepsilon_{e1}^{0} + \Sigma_{e1}^{\text{pol}} \tag{3.1}$$

where $E_1[e_1]$ is the energy of the nanoparticle after addition of one electron to the e_1 energy level, E_0 is the initial energy, ε_{e1}^{0} describes quantum confinement, Σ_{e1}^{pol} is the dielectric confinement.

The energy required for the formation of a non-interacting electron–hole pair (quasi-particle gap) (Fig. 3.1b) is given by Eq. 3.2:

$$\varepsilon_{\text{gap}}^{qp} = E_1[e_1] - E_{-1}[h_1] - 2E_0 = \varepsilon_{\text{gap}}^{0} + \Sigma_{e1}^{\text{pol}} + \Sigma_{h1}^{\text{pol}} \tag{3.2}$$

The quasi-particle gap can be determined by voltammetric experiments, and it corresponds to the electrochemical bandgap (ΔE_{el}).

The optical gap (Fig. 3.1c) is given by Eq. 3.3:

$$\Delta E_{\text{op}} = \Delta E_{\text{el}} - J_{e1,h1} \tag{3.3}$$

where $J_{e1,h1}$ is the total Coulomb interaction of the electron–hole pair. Hence, for any given quantum dot, the electrochemical energy gap is expected to be larger than the optical energy gap.

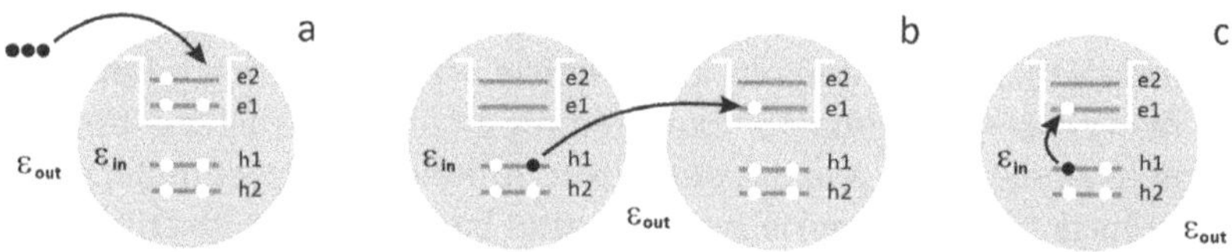

Figure 3.1 Three possible ways to charge a quantum dot: (a) electron addition to a neutral nanoparticle; (b) simultaneous injection of a hole and an electron in two non-interacting quantum dots; (c) optical excitation of an electron–hole pair. ε are the dielectric constants of the medium (ε_{out}) and of the quantum dot (ε_{in}); *e* and *h* are the energy levels of the electron and the hole, respectively. See also Colour Insert.

In the context of QD characterization, an important feature of electrochemical methods is that they mainly probe the surface of the material, whereas spectroscopic measurements are usually related to processes that take place in the interior of the nanoparticle (Fig. 3.2) [21]. Therefore, electrochemical measurements can be related to spectroscopic observations, and give information on the valence and conduction bands of the material, as long as the quality of the crystal is good [26]. It should be noted, however, that the surface of a crystal particle is inherently defective because of unsaturated valences and dangling bonds, and that surface properties are enormously important for nanoparticles due to their high surface-to-volume ratio. Moreover, defects can be generated by imperfections in the crystal lattice, the presence of impurities, and the occurrence of chemical reactions with, for example, oxygen or humidity. Surface defects can act as trap sites for both electrons and holes, thereby constituting oxidizing and reducing centers, respectively. It follows that electrochemical measurements on QDs are strongly affected by these trap states: as surface defects can hardly be controlled or cancelled, this is probably the main reason why the results of electrochemical studies performed on QDs are often contradictory and ambiguous. On the other hand, the possibility to probe the surface of quantum dots with electrochemical experiments can be exploited to gain information on the quality of the material; in this regard, electrochemical methods represent a powerful complement for the standard characterization techniques of semiconductor nanocrystals.

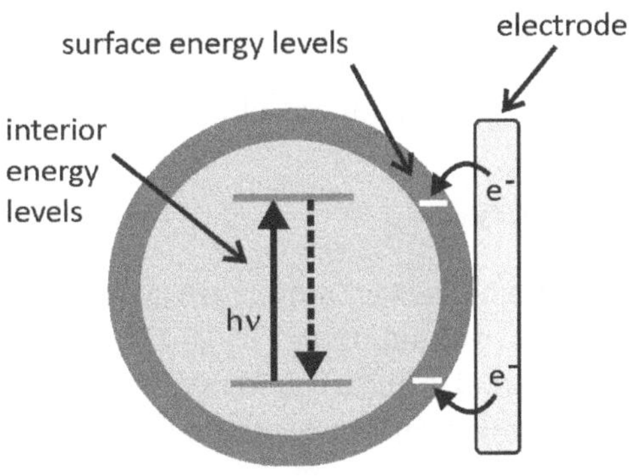

Figure 3.2 Schematic representation of the different abilities of spectroscopic and electrochemical techniques: light absorption and emission provide information about the electronic transitions in the interior of the nanoparticle, electrochemical methods mainly probe the surface of the material. See also Colour Insert.

3.3 Overview of Electrochemical Techniques Employed for the Investigations on QDs

Although a detailed description of electrochemical methods is beyond the scope of this chapter and can be found elsewhere [23], it is worth recalling some basic features of the techniques that are most commonly used for the electrochemical characterization of QDs. First of all, measurements can be performed both on solutions (or dispersions) of QDs and on thin films containing the QDs and deposited onto the electrode surface. The choice of these different setups depends on the aim of the study and on the technique employed. The main drawback of experiments in solutions is the low solubility of nanocrystals [21]: In some cases this implies that the signal is too low to be detected. The deposition of the nanocrystals onto the surface of an electrode has the advantage of providing a larger electrochemical signal: Not surprisingly, a large fraction of the studies dealing with the electrochemical properties of QDs were carried out on films. Thin films are probably the most useful for optoelectronic applications [27], but they present limitations for voltammetric experiments [28]: For example, the electron-transfer process may be followed by a chemical reaction that causes a decomposition of the film, with a consequent degradation of the sample. Moreover, as the aggregation of the nanocrystals in the film could easily take place, the observed current–potential curves may not be correlated to the behavior of individual QDs.

3.3.1 Voltammetry

Most common voltammetric techniques are cyclic voltammetry (CV) and differential pulse voltammetry (DPV) (Fig. 3.3) [23]. In a CV experiment, the potential is linearly and reversibly changed between two limits E_i and E_s with a certain scan rate (Fig. 3.3a). The resulting current–potential curve for a redox couple in solution presents two waves (Fig. 3.3b), corresponding to the anodic (oxidation) and cathodic (reduction) reactions. The average of the anodic and cathodic peak potentials (E_{pa} and E_{pc}, respectively) is called the halfwave potential ($E_{1/2}$), and in most instances coincides with the standard potential ($E°$) of the redox couple. A redox process characterized by fast electron-transfer kinetics is said to be electrochemically reversible; conversely, chemical reversibility indicates the fact that the electrogenerated species is stable within the time frame of the experiment. The features of a chemically and electrochemically reversible process are: (i) the independence of $E_{1/2}$ on the scan rate, (ii) a difference between the peak potentials of 57/

n mV at 25°C (n is the number of electrons exchanged in the redox reaction), (iii) identical anodic and cathodic peak current intensities (i_{pa} and i_{pc}, respectively), and (iv) peak current intensities proportional to the square root of the scan rate. The peak current intensity for an experiment in solution depends also on the electrode surface area, the concentration of the electroactive species, and the square root of its diffusion coefficient. For an electrochemically irreversible process (i.e., with slow electron-transfer kinetics) the CVs may or may not exhibit waves; if waves are observed, their peak potentials depend on the scan rate. The most straightforward indicator of chemical irreversibility of a redox couple (i.e., the electrogenerated species undergoes an irreversible chemical reaction) is the fact that the anodic and cathodic waves exhibit different intensities.

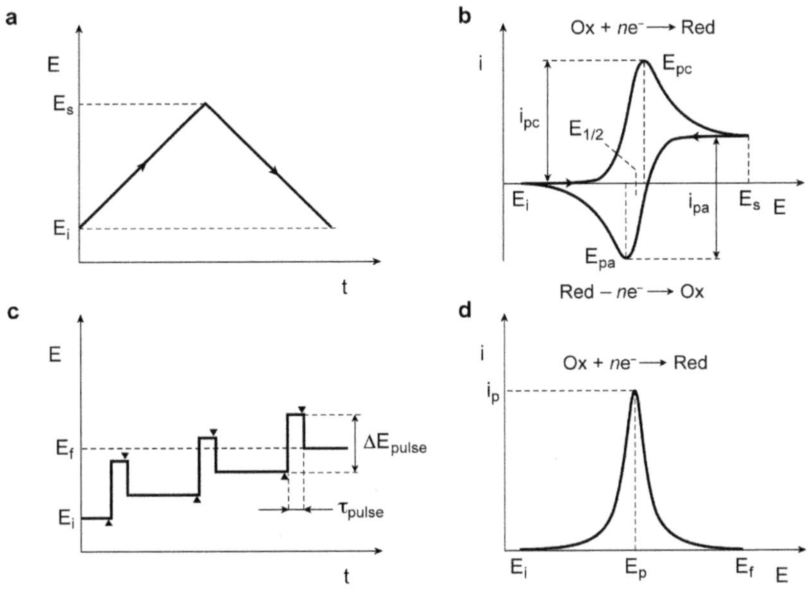

Figure 3.3 Potential profiles (a and c) and current–potential curves (b and d) for cyclic voltammetry and differential pulse voltammetry, respectively.

CV is a powerful technique because it can provide information on the energetics of the redox process, the electron-transfer rate between the electrode and the electroactive species, the diffusion coefficients, and the thermodynamic and kinetic parameters of chemical reactions possibly coupled with the redox reactions. However, the low sensitivity of CV (the detection limit in optimal conditions is around 10^{-4} M), together with the poor solubility of quantum dots in most solvents, limits the use of this technique in the investigation of the electrochemical properties of nanocrystals in solution.

The sensitivity issue can be overcome by using differential pulse voltammetry, which suppresses the background signal and enhances the sensitivity. In DPV a series of small potential pulses of a fixed height (ΔE_{pulse}) and width (τ_{pulse}) are applied over a base potential, which is increased in small increments after every pulse (Fig. 3.3c). The voltammogram is the plot of the difference between the current values measured at the end of the pulse and just before the pulse (triangles in Fig. 3.3c) versus the base potential. The differential nature of the measurement gives rise to a peaked output (Fig. 3.3d), in contrast with the wave-like appearance of CVs.

For a reversible process the DPV peak potential is independent on the scan rate, and is related to the halfwave potential determined from CVs by a simple equation [23]. The peak current depends on the electrode surface area, the concentration of the electroactive species, and the square root of its diffusion coefficient — as for CV — but also on the pulse height and duration. Specifically, the current increases on increasing the pulse height; however, also the width of the peak increases, thereby limiting the capability of the technique to resolve peaks that occur at similar potential values. Hence, a trade-off between sensitivity and resolution should guide the choice of the pulse height. Under optimal conditions, DPV allows the detection of electroactive species at concentrations as low as 10^{-8} M. In the case of chemically or electrochemically irreversible processes, the DPV peaks tend to become broader and weaker, and their potential becomes dependent on scan rate.

3.3.2 Electrochemiluminescence

Electrochemically generated chemiluminescence (electrochemiluminescence, ECL) is commonly defined as the emission of light resulting from the generation of electronically excited states formed in homogenous electron transfer reactions between active precursors obtained by means of heterogeneous electron transfer processes [23, 29]. The reaction sequence to generate an excited state and light emission is

$$A + e^- \rightarrow A^- \tag{3.4a}$$

$$D - e^- \rightarrow D^+ \tag{3.4b}$$

$$A^- + D^+ \rightarrow {}^*A + D \tag{3.4c}$$

$$^*A \rightarrow A + h\nu \tag{3.4d}$$

where A and D could also be the same species. An alternative approach for the production of electrochemiluminescence is the use of a coreactant: for example an electroactive species O that, after reduction, produces a strong oxidizing agent O^+, which can react with the anion A^-:

$$O + e^- \rightarrow O^- \tag{3.5a}$$

$$O^- \rightarrow O^+ + \text{products} \tag{3.5b}$$

$$A^- + O^+ \rightarrow {}^*A + \text{products} \tag{3.5c}$$

$$^*A \rightarrow A + h\nu \tag{3.5d}$$

Similarly for the anodic side,

$$R \rightarrow R^+ + e^- \tag{3.6a}$$

$$R^+ \rightarrow R^- + \text{products} \tag{3.6b}$$

$$D^+ + R^- \rightarrow {}^*D + \text{products} \tag{3.6c}$$

$$^*D \rightarrow D + h\nu' \tag{3.6d}$$

The use of a coreactant is common in the ECL studies of semiconductor nanocrystals, whose electrogenerated species are usually unstable. ECL revealed to be very useful for the investigation of QDs because it correlates electrochemical and photophysical processes [21]. As the narrow emission band is a peculiar signature of quantum dots, the ECL spectrum can facilitate the assignment of the electrochemical processes.

3.3.3 Spectroelectrochemistry

Non-electrochemical techniques, such as absorption and emission spectroscopy, can be coupled with an electrochemical system in order to study the spectroscopic behavior of electrogenerated species [23]. Most spectroelectrochemical setups are based on transmission measurements, in which a light beam is sent through a transparent electrode surface, and the absorbance changes resulting from species produced or consumed in the electrochemical reaction are monitored. In other instances, it is the photoluminescence that is measured during the electrochemical experiment. These investigations involve the study of the absorbance or the luminescence intensity as a function of time as the electrode potential is changed (e.g., in a controlled potential electrolysis); the spectral signature of the electrogenerated species can also be determined by means of a wavelength scan.

An important issue in spectroelectrochemistry (SEC) experiments is the timescale of the electrolysis, which should be ideally fast enough to permit the spectroscopic detection of electrogenerated species before they become involved in chemical reactions. Electrolysis times can be shortened using electrodes with large areas and cells with small solution volumes, such as in thin-layer cells.

3.4 CdSe Nanocrystals

CdSe quantum dots are probably the most extensively investigated class of II–VI semiconductor nanocrystals, owing to their valuable optical properties (especially if coated with a ZnS shell) and the availability of reliable synthetic procedures — although the toxicity of cadmium poses concerns for their practical applications. A large amount of data on their synthesis, characterization, and properties are available in the literature [7, 30–35].

Several groups investigated the electrochemical behavior of CdSe and CdSe–ZnS core–shell nanocrystals [21]. The studies involve cyclic voltammetry, both in dispersed solution and on thin films, differential pulse voltammetry, SEC, and electrochemiluminescence.

3.4.1 Voltammetry

CV measurements on CdSe nanocrystals in solution and on films typically reveal a good match between optical and electrochemical bandgaps [28, 36, 37]. Haram and coworkers [28] reported a systematic study on a series of CdSe quantum dots capped with tris-*n*-octylphosphine oxide (TOPO) ligands, with sizes ranging from 1.9 to 3.6 nm. They observed the presence of one anodic and one cathodic irreversible waves in the CVs (Fig. 3.4), respectively assigned to the removal of one electron from the valence band and injection of one electron into the conduction band. The difference between the anodic and cathodic peak potentials increases on decreasing the size, as expected from the quantum confinement of the charges. A comparison of the data from three different papers [28, 36, 37] revealed a satisfactory agreement between independent measurements as well as with the values predicted theoretically (Fig. 3.5) [24]. The energy associated with the CV peak-to-peak separation, i.e., the quasi-particle gap (Section 3.2), is in good agreement with the optical bandgap (Fig. 3.6), with the only exception of the smaller particles (<1.9 nm). In this case, as expected from the theory [24], the optical bandgap is not equal to the quasi particle gap because of the Coulombic electron–hole interaction term (Eq. 3.3), $J_{e1,h1}$: based on this model, the value of $J_{e1,h1}$ for the smaller particles was estimated to be *ca.* 0.1 eV. The QDs with a diameter of 1.9 nm differ from the larger samples also for the presence of a further anodic peak in the CVs, which is assigned to inter-band trap states, as also evidenced from the presence of a large shoulder on the low-energy side of the photoluminescence band.

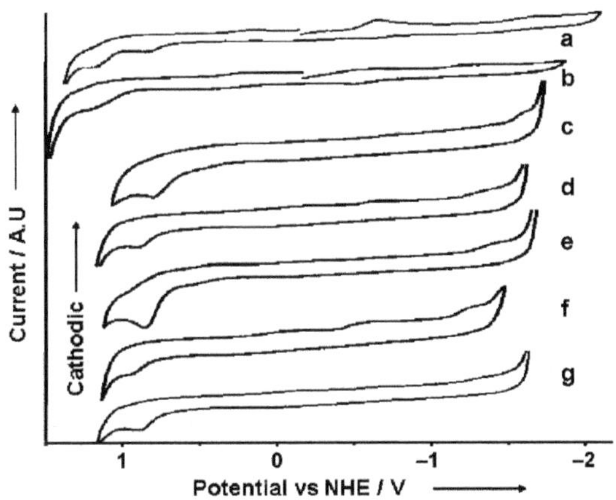

Figure 3.4 Cyclic voltammograms of DMSO-toluene solutions of TOPO-capped CDSe quantum dots (1 mg/mL) of increasing diameter size from **a** (1.9 nm) to **g** (3.6 nm). From ref. 28.

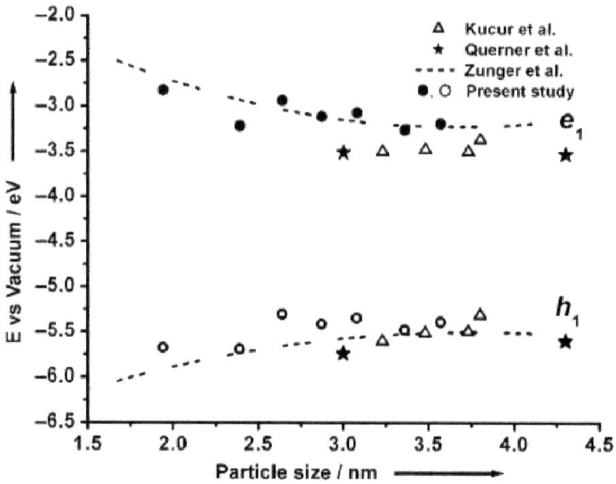

Figure 3.5 Energy values of conduction and valence bands of CdSe quantum dots determined from electrochemical experiments [28, 36, 37] and theoretical calculations [24]. From ref. 28.

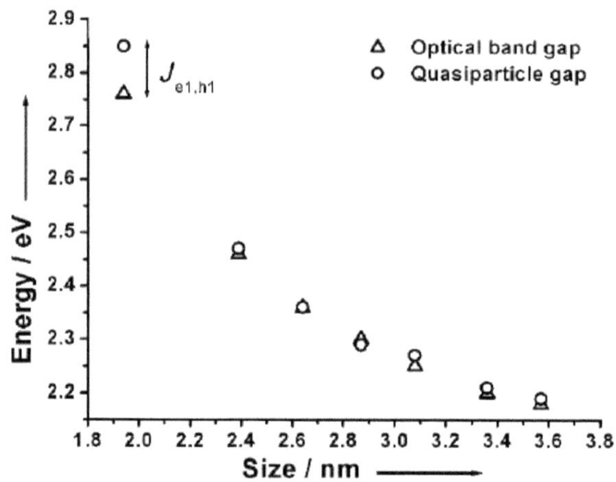

Figure 3.6 Comparison of optical bandgap (determined from UV/V is spectra) and quasi-particle gap (determined from voltammetric measurements) of CdSe quantum dots. From ref. 28.

The presence of additional peaks in the voltammograms of quantum dots, besides the processes associated with oxidation and reduction of the conduction and valence bands, is quite common. Indeed, electrochemical techniques explore the surface of the material and are sensitive to the presence of defects, which are responsible for the presence of oxidizing or reducing trap states. In this regard, Raymo, Credi, and coworkers [38, 39] investigated three series of CdSe core quantum dots, prepared with different synthetic procedures. The resulting materials were coated with different organic ligands, namely *n*-decanethiol (**a1–a3**), TOPO and *n*-hexadecylamine (HDA) (**b1–b4** and **c1–c3**). The electrochemical properties of these nanocrystals were investigated by means of differential pulse voltammetry. In all cases, the DPV scans show one oxidation and one reduction processes. The electrochemical bandgap energies of the *n*-decanethiol-coated samples (Table 3.1, Fig. 3.7), decrease on increasing the core diameter, as expected from the theory. Nevertheless, the electrochemical bandgap is always 0.4–0.5 eV smaller than the optical bandgap. This discrepancy can be ascribed to the presence of surface defects that act as local trap states for electrons and holes. Such an interpretation is supported by the presence of a tail in the emission band of this set of quantum dots.

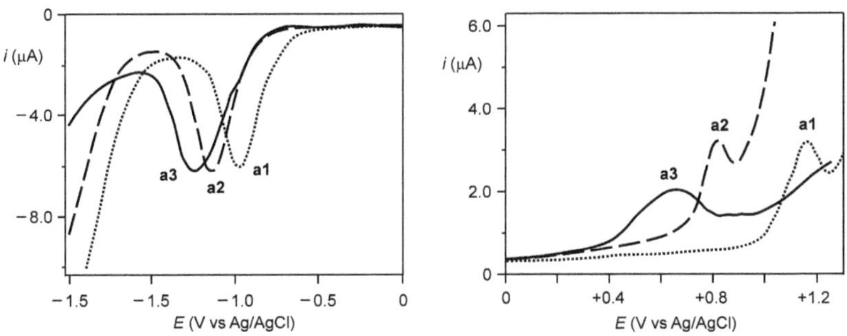

Figure 3.7 Differential pulse voltammograms for reduction (left) and oxidation (right) of THF solutions of *n*-decanethiol coated CdSe quantum dots (**a1–a3** series, 1.5 μM) [48].

The TOPO-HDA coated samples were prepared in two solvents exhibiting different reactivity towards the surface of the nanocrystals, namely 1-*n*-octadecene (samples **b1–b4**) and TOPO (samples **c1–c3**). Octadecene (ODE) behaves as a non-coordinating solvent, while TOPO is an efficient capping ligand for QDs. Interestingly, for these samples the redox potentials were not monotonously related to the core size (Table 3.1, Figs. 3.8 and 3.9). In particular, the electrochemical bandgap increases on increasing the diameter for samples **b1–b3** and **c1–c2**, in contrast with what expected from the theory and from the optical bandgap. Moreover ΔE_{el} is smaller than ΔE_{op}, (contrary to what expected on the basis of Eq. 3.3) for the smallest and largest samples of each series (**b1, b4, c1, c3**). It is likely that surface defects play a major role in determining the redox potentials, and the influence of the defects is dominant for the smallest and largest particles of these particular batches of quantum dots.

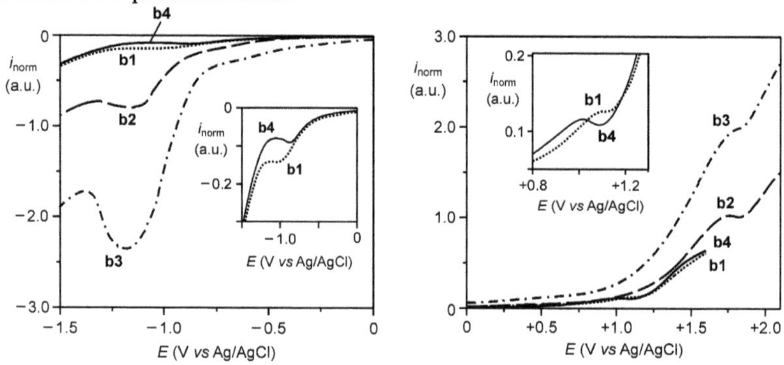

Figure 3.8 Differential pulse voltammograms for reduction (left) and oxidation (right) of $CHCl_3$ solutions of TOPO–HDA coated CdSe quantum dots (**b1–b4** series, 1.5 ÷ 4 μM) [39]. The DPV peaks of **b1** and **b4** are magnified in the insets.

Table 3.1 Redox potentials and electrochemical and optical bandgaps of CdSe quantum dots coated with different organic ligands and with different core diameters

Sample	d_{core} (nm)[a]	E_{ox} (V)[b]	E_{red} (V)[b]	ΔE_{el} (V)	ΔE_{op} (V)
a1	2.1	+1.16	−1.07	2.24	2.66
a2	2.3	+0.82	−1.20	2.03	2.53
a3	2.5	+0.68	−1.26	1.94	2.41
b1	2.6	+1.10	−0.99	2.09	2.30
b2	3.1	+1.79	−1. 21	3.00	2.18
b3	3.8	+1.77	−1.25	3.02	2.09
b4	4.3	+1.02	−0.89	1.91	2.04
c1	2.1	+1.14	−0.86	2.00	2.41
c2	4.1	+1.32	−1.18	2.50	2.04
c3	4.6	+1.24	−0.82	2.06	2.01

Source: Data from refs. 39 and 48.

[a]Determined from the position of the low-energy absorption peak.

[b]Potential values for the first oxidation and reduction, in V vs. Ag/AgCl (KCl 3M), determined from the DPV peak.

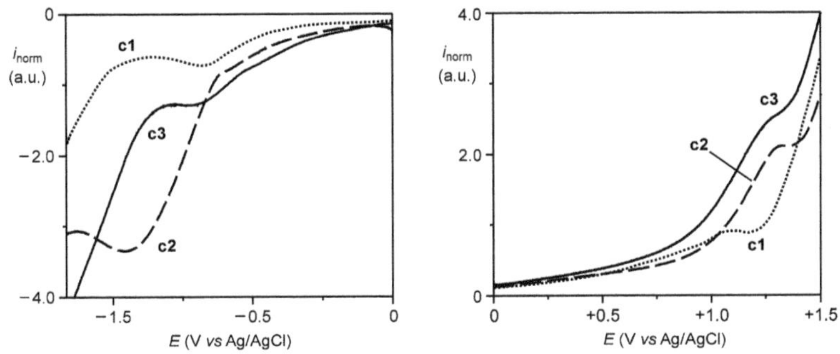

Figure 3.9 Differential pulse voltammograms for reduction (left) and oxidation (right) of $CHCl_3$ solutions of TOPO–HDA coated CdSe quantum dots (**c1**–**c3** series, 3 μM) [39].

A possible explanation of this behavior takes into account the methodology employed for the synthesis of these nanocrystals, that is, the hot injection method [40]. Small particles have more defects because of the relatively short reaction time which does not allow thermal annealing. Mid-sized particles, growing in the focusing regime, most likely experience a correction of surface defects besides the narrowing of the size distribution. The growth of larger particles can occur, at least in part, by Ostwald ripening (disassembly of small crystals and incorporation of the resulting material into larger ones) [41]. Therefore, the building blocks reacting on the surface in this regime derives from small QDs, which carry a substantial amount of defects.

Moreover, mid-sized particles of each series (**b2**, **b3**, **c2**) show considerably higher potentials for reduction and oxidation than the other particles of the same family, and exhibit an electrochemical gap much larger than that expected on the basis of their optical bandgap (Eq. 3.3). It should be noted that the DPV peaks for samples **b2**, **b3** and **c2** are quite broad. This observation would be easily accounted for by the presence of a broad size distribution, but this is not the case for the present samples. Reasonable possibilities are that (i) the DPV peak is in fact the result of several electron-transfer processes occurring at slightly different potentials, and (ii) the electron exchange between the electrode and the nanocrystals exhibits slow kinetics. The first interpretation is in line with earlier observations that quantum dots can undergo multielectron redox processes [42–45], and is consistent with the relatively high current intensities associated with the DPV peaks. Regarding the second possibility, nanocrystals with low surface defect rates could indeed undergo sluggish electron-transfer reactions because their surface is more effectively coated by the electrically insulating organic molecules (defected sites may not act as binding sites for the ligands) [46], and the redox processes could involve atoms located in the nanocrystal interior.

Quantum confinement effects indicate that the oxidation and reduction processes should become more difficult as the size of the particle decreases. Actually, even though the electrochemical band-gap is in agreement with the theoretical expectation in some cases, the absolute values of the potentials for oxidation and reduction do not seem to follow any particular trend.

An inorganic shell of a second semiconductor material on the top of the nanocrystal core removes the discontinuity represented by the core–ligand interface, partially eliminates surface defects and protects the core from the interaction with the environment [47]. These core–shell quantum dots usually present improved physicochemical properties and stability. The electrochemical behavior of core–shell CdSe–ZnS quantum dots with different shell thicknesses was investigated by Raymo, Credi, and coworkers [48]. Differential pulse voltammograms of 2.2 nm CdSe cores with ZnS shell thicknesses of 1.4, 2.2 and 5.0 nm show one oxidation and one reduction processes in the investigated potential window. The presence of the shell shifts both potentials in the positive direction with respect to the core, thus discouraging oxidation and slightly facilitating reduction. The positive potential shift of the oxidative process is in qualitative agreement with the change in the valence band energy on going from CdSe to ZnS [49], suggesting that the oxidation of core–shell nanoparticles may involve the removal of electrons from the inorganic shell.

3.4.2 Electrochemiluminescence

As noted above, electrochemiluminescence is a valuable technique to investigate the nature of the electroactive species. It was found that CdSe nanocrystals with a diameter of 3.2 nm emit light through the annihilation mechanism (Eq. 3.4), on scanning the potential from +2.3 V to −2.3 V versus an Ag pseudoreference electrode (Fig. 3.10) [50–51]. The CV of the solution does not show any significant feature, probably because of the low solubility of the sample in the CH_2Cl_2 solvent. From the relative intensity of the ECL emission in the oxidation and reduction cycles, it can be inferred that the electrogenerated oxidized forms are more stable than the reduced ones. Surprisingly, ECL could be observed on scanning the potential both in negative (0 ÷ −2.3 V) and positive (0 ÷ +2.3 V) directions [50], a phenomenon which was ascribed to the presence of impurities.

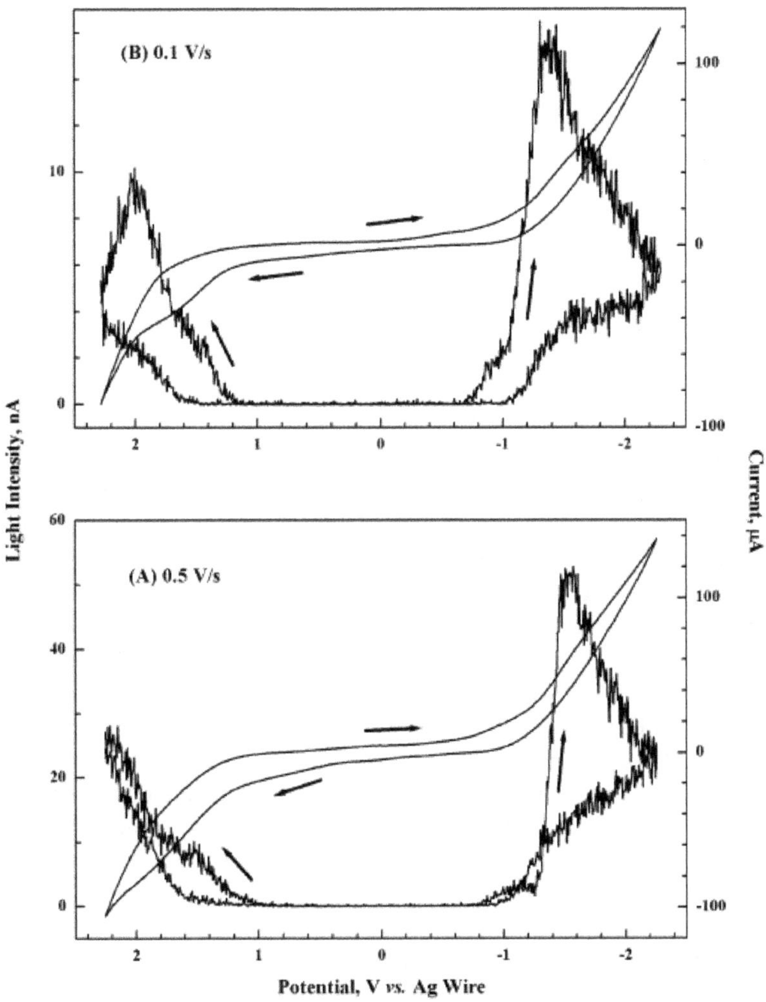

Figure 3.10 Cyclic voltammograms and electrochemiluminescence curves of CH_2Cl_2 solutions of CdSe quantum dots at different scan rates. From ref. 50.

The same behavior was observed on 2.5 nm CdSe nanocrystals on scanning the potential between 0 and −3.00 V [51], versus an Ag pseudoreference electrode. Stable ECL was observed even at pseudo constant potentials on the negative side, slowly scanning the potential from −1.60 to −1.80 V or from −2.00 to −2.20 V. The interpretation of these data involves a coreactant mechanism (Eq. 3.5), wherein the ECL is emitted from the excited states formed by the reaction of the anion radicals with oxygen or solvent molecules. The ECL spectrum obtained on scanning the potential in the negative direction presents two broad bands (653 nm and 486 nm) attributed to emission involving the surface states and the core structure, respectively.

In the case of the 3.2 nm CdSe nanocrystals discussed previously [50], the ECL spectrum obtained on scanning the potential between +2.3 and −2.3 V exhibits a maximum at 740 nm, with a red-shift of 200 nm with respect to the photoluminescence maximum. This difference is most likely the result of a major contribution of the surface states in ECL emission, and is expected to be reduced by passivation of the nanocrystal surface [47]. Indeed, the construction of a CdSe–ZnSe core–shell structure resulted in a dramatic enhancement of photoluminescence [52]. The ECL spectrum of these core–shell nanocrystals (Fig. 3.11) [52] shows one strong band at 580 nm, which is almost identical to that observed in the photoluminescence spectrum, and a second broad band at 740 nm, i.e., strongly red shifted from the photoluminescence peak. This spectrum suggests that the solution contained nanocrystals whose surface had been efficiently passivated, together with non-passivated particles.

The ability to probe the surface states of nanocrystals is an important advantage of ECL. The fact that different results are obtained with the same samples suggests that this technique can be exploited to gather information on the quality of the nanocrystals.

In order to obtain intense and efficient electrogenerated chemiluminescence, a coreactant mechanism (Eqs. 3.5 and 3.6) can be used. Talapin and coworkers [53] prepared close-packed thin films of CdSe nanocrystals on platinum or F-doped SnO_2 substrates, and studied their electrochemical properties in 0.1 M $K_2S_2O_8$ aqueous solution. The disulfate anions can be reduced, producing sulfate radicals (Eq. 3.7) that inject holes into the valence band of the electrochemically reduced semiconductor film (Eq. 3.8) [54]. A subsequent fast recombination of the electron–hole pair produces emission of light (Eq. 3.9).

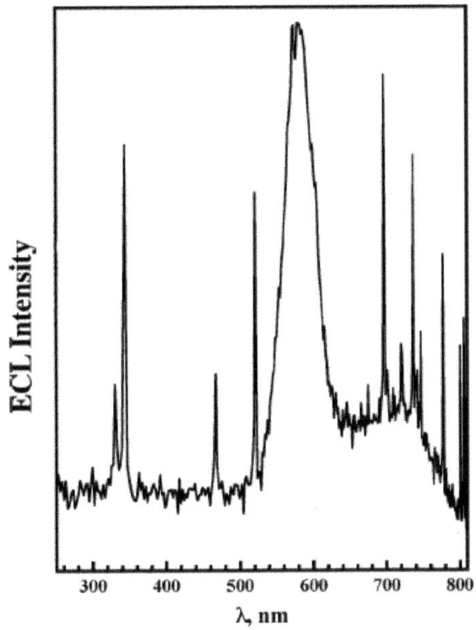

Figure 3.11 Electrochemiluminescence spectrum of a CH_2Cl_2 solution of CdSe/ZnSe core/shell quantum dots, obtained by potential steps between +2.3 and –2.3 V. From ref. 52.

$$CdSe + e^- \rightarrow CdSe^- \quad (3.7a)$$

$$S_2O_8^{2-} + e^- \rightarrow SO_4^{2-} + SO_4^{\cdot-} \quad (3.7b)$$

$$SO_4^{\cdot-} + CdSe^- \rightarrow SO_4^{2-} + {}^*CdSe \quad (3.8)$$

$$^*CdSe \rightarrow CdSe + h\nu \quad (3.9)$$

The ECL spectra of nanocrystals with different diameters show an emission band with the same shape and width as that observed in the photoluminescence spectra of the films, but slightly red-shifted because of local fluctuations of packing density inside the nanocrystal film. The ECL spectrum of films made with *ca.* 3 nm CdSe quantum dots exhibited, beside the sharp band associated to the exciton recombination, a broad band assigned to trap-state emission (Fig. 3.12).

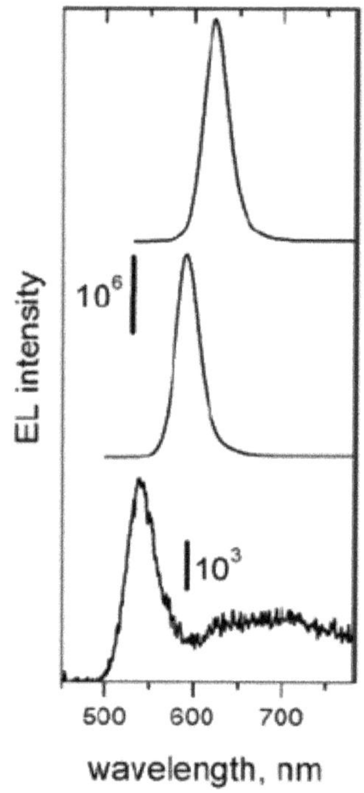

Figure 3.12 Electrochemiluminescence spectra of close-packed thin films of CdSe quantum dots of 5.1, 4.2, and 2.8 nm (from top to bottom). From ref. 53.

ECL presents several advantages with respect to photoluminescence as a sensing technique [29], and examples of ECL sensors based on quantum dots have been reported [55].

3.4.3 Spectroelectrochemical Measurements

Guyot-Sionnest and coworkers [27, 56] investigated CdSe and CdSe–ZnS nanocrystals films and solutions, and monitored their absorption and luminescence spectra upon application of an electrochemical potential. Under a negative potential, a new absorption band in the near infrared region is observed, which was assigned to the $1S_e \rightarrow 1P_e$ transition; concomitantly, the exciton absorption band is bleached. Moreover, the luminescence of the negatively charged nanocrystals is quenched. The reversibility of the charging process was verified by following the reset of both the absorption and luminescence signals. Interestingly, the authors observed different reset timescales for absorption and emission [56]; in particular, the photoluminescence recovery rate was slower in comparison with the IR absorption decay. This effect was attributed to the presence of surface charges: the IR absorption is due to the electrons in the 1S state, whereas the photoluminescence is strongly quenched by charges in the surface bandgap states. Overall, the recovery process was slow, and could be improved treating the films with cross-linkers [27].

Some basic phenomena pertaining to the photophysical behavior of semiconductor nanocrystals, such as blinking and photobrightening, are still under investigation. As tentative explanations of these phenomena often involve the participation of charges, insightful information can be inferred from emission studies on charged nanocrystals. For this purpose, Mulvaney and coworkers [46] combined cyclic voltammetry and confocal microscopy to study 2D monolayers of CdSe–CdS–ZnS semiconductor nanocrystals. The experiments, performed in air and under nitrogen, showed that the luminescence of the nanocrystals is completely and irreversibly quenched when the potential is scanned in the positive direction (Fig. 3.13a), both in air and under N_2. Upon scanning the potential in the negative direction, the photoluminescence is partially and reversibly quenched under N_2 (Fig. 3.13b), whereas it is substantially and almost reversibly enhanced in air (Fig. 3.13c). In any case, the electron injection occurs at potentials well below the predicted conduction band energy for bulk ZnS; therefore, it appears that the 1.05 nm-thick inorganic shell does not hamper the injection of charges in the CdSe core. The oxidation of the nanocrystals probably causes chemical decomposition: as there was no evidence of dissolution, it is likely that the degradation occurs at the internal CdSe–ZnS interface. The behavior observed in response to electron injection is strongly dependent on the environment, and can be accounted for by the mechanism shown in Fig. 3.14. In deaerated conditions, electrons are injected into the unoccupied $1S_e$ level (step 1) and rapidly relax to surface acceptor sites (step 2). Such trap sites are located at the ZnS shell surface and cause the quenching of the nanocrystal photoluminescence, because the trapped electron recombines non-radiatively with photogenerated holes (step 3), and the residual excited electron is retrapped at the vacant surface state (step 4). In aerated solution, an initial slight quenching of the photoluminescence (ascribed to the surface trapping discussed above) is followed by photobrightening of the nanocrystals. The mechanism proposed to interpret such a behavior involves a role of the water molecules that are always present in an air-equilibrated organic solvent. Briefly, water can strongly chemisorb at the surface trap sites when the latter become vacant, i.e., when the trapped electron recombines with a hole generated by photoexcitation. The binding of H_2O is in competition with the transfer of the photoexcited electron from the CdSe core to the vacant trap site (step 4 in Fig. 3.14). Such a "photoannealing" process causes a progressive passivation of the trap sites which mediate non-radiative recombination (see above), thereby leading to an increase in the photoluminescence quantum yield upon light irradiation [57].

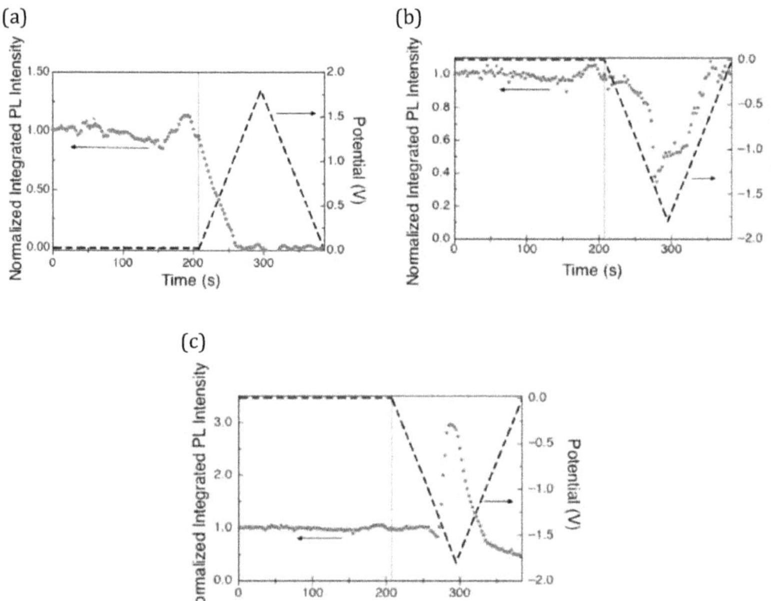

Figure 3.13 Integrated photoluminescence intensity (red circles) and applied potential (black dashed line) as a function of time for a monolayer of CdSe/CdS/ZnS quantum dots (a) during positive charging in dry air and under negative potential (b) under dry nitrogen and (c) in air. From ref. 46.

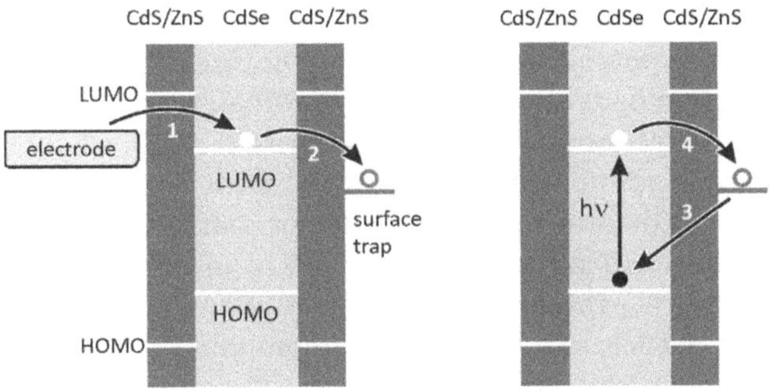

Figure 3.14 Mechanism for the quenching of the photoluminescence from negatively charged quantum dots in nitrogen: injection of an electron into the LUMO level (step 1) and rapid relaxation to the empty surface trap (step 2); recombination of the trapped electron with the photogenerated hole (step 3), followed by trapping of the photoelectron in the surface state (step 4). From ref. 46.

3.5 CdTe Nanocrystals

3.5.1 Voltammetric Measurements

Cyclic and differential pulse voltammetric experiments on CdTe nanocrystals confirm what generally observed for CdSe quantum dots: the electrochemical response is strongly influenced by the quality of the material. Bard and coworkers [58] observed several oxidation and reduction peaks in the differential pulse voltammetry of two batches of TOPO-capped CdTe. The most intense oxidation peak was assigned to oxidation of the Te ions, while the other processes were ascribed to impurities. The Te oxidation, however, was sometimes not distinguishable from the background. The determination of the diffusion coefficient from the DPV peaks gave somewhat contradictory results.

Talapin and coworkers [59] performed a systematic investigation on several batches of thiol-capped CdTe nanoparticles of different sizes in aqueous solution and deposited on an electrode. The cyclic voltammograms of the CdTe samples (Fig. 3.15) show two oxidation peaks. The first one (A_1) shifts to more positive potentials on increasing the size (i.e., on decreasing the bandgap) of the nanoparticles, whereas the second oxidation peak (A_2) moves slightly to less positive potentials. The scan on reduction, which is not shown in Fig. 3.13, shows one irreversible peak (C_1) that shifts to more negative potentials on decreasing the size of the nanocrystals. The separation between A_1 and C_1 is too small to correlate with the bandgap energy; moreover, the potential shift of A_1 as a function of the QD size is not in agreement with quantum confinement effects. This peak was therefore assigned to the oxidation of surface defects forming intra-bandgap surface states. This hypothesis was confirmed by voltammetric experiments on samples of quantum dots with the same mean particle size and different photoluminescent efficiencies. Weakly luminescent nanoparticles have the same energy gap as the strongly luminescent ones but more Te atoms on their surface that can act as traps [60]. The CV of these samples showed that the charge transferred in correspondence of peak A_1 is larger for samples characterized by lower photoluminescence yields (i.e.,

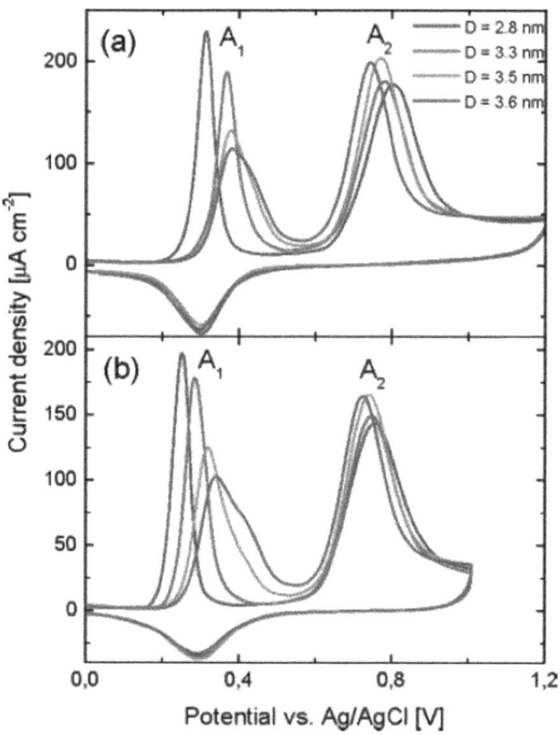

Figure 3.15 Cyclic voltammograms of thiol-coated CdTe quantum dots of different sizes (a) in buffer solution (a) or adsorbed on a gold electrode (b). From ref. 59.

by a higher number of traps), thus confirming the nature of the redox process. The second oxidation peak A_2 was then assigned to the oxidation of the CdTe core; the bandgap energy was calculated as the energy difference between A_2 and C_1. The electrochemical bandgap was found to be smaller than the optical bandgap, a fact that was attributed to a multielectron transfer process followed by decomposition of the particles [61].

A systematic study on the electrochemical response of both CdSe and CdTe quantum dots coated with most commonly employed ligands, such as octadecylamine (ODA), dodecylthiol (DT) and mercaptopropionic acid (MPA) was recently carried out by Mews and coworkers [62]. The employed ligands are known to affect the luminescence intensity of CdSe QDs in an opposite manner compared with CdTe QDs [63]. Very interestingly, the authors rationalized such an effect in terms of the valence band energy of the nanocrystal with respect to the HOMO levels of the ligand. Both values were evaluated by means of cyclic voltammetry, in order to take into account possible surface effects on the energy levels. It was found that when the HOMO of the ligand lies above the valence band of the nanocrystal the luminescence is quenched because of a hole transfer from the valence band of the quantum dot to the HOMO of the ligand. These results enable the evaluation of the relative energies associated with oxidation of the nanocrystal and the surface-bound ligands from the QD luminescence intensity.

3.5.2 Electrochemiluminescence

ECL from CdTe nanocrystals in CH_2Cl_2 solution was observed on scanning the potential in the negative direction (from 0 to –2.46 V versus the saturated calomel electrode (SCE)) [58]. This signal cannot be accounted for by annihilation of radical species of the nanoparticles; its origin was interpreted in terms of a mechanism in which the solvent acts as coreactant through its reduced form:

$$CH_2Cl_2 + e^- \rightarrow CH_2Cl_2^{\cdot -} \rightarrow CH_2Cl^{\cdot} + Cl^- \quad (3.10)$$

$$CdTe^- + CH_2Cl^{\cdot} \rightarrow {}^*CdTe + CH_2Cl^- \quad (3.11)$$

The ECL spectrum shows a peak in correspondence of the photoluminescence peak (Fig. 3.16), indicating that the electrogenerated chemiluminescence arises from the passivated CdTe. The CdTe nanoparticles exhibit a relatively low photoluminescence quantum yield, in spite of their superior degree of passivation compared with CdSe core QDs [50]. This behavior is explained considering that tellurium atoms react with oxygen more easily than selenium atoms, and that the oxidized Te surface may exert a luminescence quenching effect.

The electrogenerated chemiluminescence of thiol-capped CdTe nanocrystals was investigated in aqueous solution [64]. ECL emission was observed on scanning the potential from 0 to −2.4 V versus an Ag/AgCl electrode; the proposed reaction mechanism involves reduced oxygen as a coreactant. In this case, the ECL spectrum is red-shifted with respect to the photoluminescence spectrum, indicating that trap states are generated which undergo radiative deactivation. Experiments performed on CdTe quantum dots with different diameters showed an enhancement of the ECL signal on increasing the dimension of the nanocrystals [65]. Presumably, the electrochemical generation of excitons in larger QDs is more efficient because of a better match between the energy available in the electron-transfer reaction of the electrogenerated partners and the bandgap energy. Moreover, in this particular series of QDs larger nanocrystals exhibited higher luminescence quantum yields than the smaller ones.

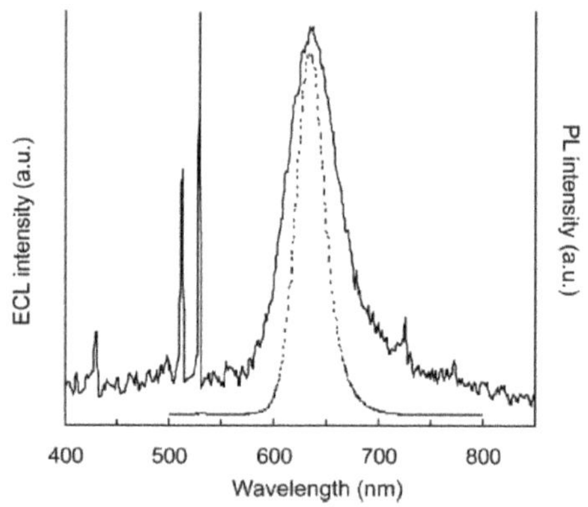

Figure 3.16 Electrochemiluminescence spectrum of a CH_2Cl_2 solution of CdTe quantum dots (7 μM), obtained by potential steps between −2.3 and +2.3 V. Dotted line is the photoluminescence spectrum. From ref. 58.

The ECL of CdTe nanoparticles, obtained by exploiting cathodic or anodic coreactant mechanisms, has been exploited for sensing applications [66–68].

3.5.3 Spectroelectrochemical Measurements

In analogy with CdSe nanocrystals (Section 3.4.3), the changes in the photoluminescence properties of CdTe quantum dots as a function of the applied electric potential were investigated both in air and under nitrogen. Multilayer films of CdTe nanocrystals and poly(diallyldimethylammonium chloride) (PDDA) were deposited on an indium tin oxide electrode, and their electrochemical behavior was investigated in aqueous solution [69]. On application of an increasing positive potential, the luminescence of the quantum dots was blue shifted and irreversibly quenched both in air and in N_2, as previously observed [46] for CdSe nanocrystals. X-Ray photoelectron spectroscopic (XPS) studies indicated that, on oxidation, a large amount of Te^0 atoms appear at the surface of the particles, acting as traps to decrease the photoluminescence efficiency [60]. While the photoluminescence quenching can thus be related to the change of the surface structure of the nanocrystals, the blue-shift suggests that the particles undergo dissolution with a subsequent decrease of their dimensions. As for the CdSe nanocrystals [46], the effect of the negative potential depends on the environment, but with different results for CdTe quantum dots: In air the luminescence is quenched, while it is not affected under nitrogen. XPS Experiments on applying a negative potential in air show again the appearance of Te^0 on the surface of the nanoparticles, which is responsible for the quenching of the luminescence [60]. The presence of Te^0 was postulated to arise from surface degradation of the colloidal nanocrystals caused by the hydrogen peroxide generated from the reduction of oxygen in aerated environment. The differences in the reduction behavior between the CdTe and CdSe nanocrystals is possibly due to the different solvents and potential windows employed in the various experiments.

3.6 Nanocrystals Functionalized with Electroactive Molecules

Most synthetic procedures yield QDs capped with molecular layers (e.g., TOPO, TOPO/HDA, aliphatic thiols) that exhibit electrically insulating behavior. Conversely, the preparation of hybrids composed of semiconductor nanocrystals and electroactive molecular ligands opens up the possibility of addressing the optoelectronic properties of the QDs by taking advantage of the redox properties of the surface ligands. Interesting examples of quantum dots functionalized with electroactive moieties have been reported [70].

Ferrocene is a simple molecular unit characterized by an excellent and well characterized electrochemical behavior; it undergoes a facile one-electron reversible oxidation to the ferrocenium cation. It has been shown that ferrocene is able to quench the luminescence of quantum dots, most likely because of a photoinduced electron-transfer process in which ferrocene plays the role of the electron donor [48, 71]. Vancso and colleagues [72] employed a modified ferrocene moiety (6-ferrocenyl-1-hexanethiol) to modify the QD surface [71b]. The nanohybrids were prepared by refluxing TOPO-coated CdSe–ZnS nanocrystals with a large excess of the thiol-terminated ferrocene ligand. The product was purified by ultracentrifugation to ensure the removal of any free ferrocene molecule. The electrochemical experiments carried out on the nanohybrid showed that the electrochemical response of both the QD and the ferrocene units is changed with respect to the separated components (i.e., the native TOPO-capped QDs and bare ferrocene). Specifically, the nanohybrids exhibit a CV wave at +0.36 V versus the standard hydrogen electrode (SHE) electrode, assigned to oxidation of the surface-bound ferrocene units. This process occurs at a lower potential value compared with free ferrocene (+0.51 V), indicating that incorporation into the ligand shell causes a stabilization of the ferrocenium cation relatively to the neutral form. As far as the nanocrystal-centered processes are concerned, the oxidation peak corresponding to the removal of an electron from the QD highest occupied level is almost completely suppressed, indicating that the positively charged ferrocene unit at the QD surface hampers substantially the extraction of an electron from the nanocrystal. It is likely that the anodic peak is shifted to more positive values, falling in a potential region which is not accessible under the employed experimental conditions. Moreover, the QD reduction peak is shifted to more negative values with respect to the TOPO-capped nanocrystals.

The functionalization of bare CdSe core quantum dots with redox active oligoaniline ligands was described by Querner *et al.* [37]. The electrochemical behavior of the resulting nanoconjugates was investigated and compared with that of TOPO-coated QDs. The surface functionalization was achieved by a two-step procedure.

The first step involves the exchange of the native TOPO with 4-formyldithiobenzoic acid ligands. In the second step, an aniline tetramer is grafted via a condensation reaction between the aldehyde group of the ligand and the primary amine group of the tetramer. The presence of aniline tetramers at the QD surface strongly affects the voltammetric response of the nanocrystals. The aniline-QD nanohybrids show a new reversible process in a region where the TOPO-coated QDs do not show any electrochemical activity. Such a new process was thus attributed to the organic part of the nanohybrid, and specifically to the oxidation of the amine groups in the surface ligands. The QD component of the nanoconjugates becomes more difficult both to oxidize and to reduce. The positive shift of the potential for oxidation was attributed to the electrostatic effect exerted by the positive charging of the aniline ligands prior to the QD oxidation. The shift of the reduction wave towards more negative reduction was instead related to the insulating nature of the ligands. The nanohybrids were also investigated by SEC, looking for the spectral signature of the oxidized and reduced forms of the organic ligand. Going toward the oxidation side, the spectral features indicate a partial oxidation of the ligand, which was already observed to a lesser extent when the potential was not applied. The bands of the ligand in its neutral form disappear, while a new broad band typical of a polaron (radical cation) charge carrier shows up in the near IR region. All these results show that the ligands influence the electrochemical response of the nanocrystals and *vice versa*, thereby indicating a substantial degree of electronic communication between the organic and inorganic components of these nanohybrids. Finally, the possibility of an electrochemical switching of the electroactive ligand between its non-conducting (aniline neutral form) and conducting (radical cation) states was demonstrated; this property is not exhibited by the QDs coated with the native electrochemically inactive TOPO ligand.

In order to develop ion-sensitive nanohybrids, CdSe and CdTe nanocrystals were functionalized with the crown ether 4,7-di(1-aza-4,7,10,13,16-pentaoxacylooctadecyl)1,10-phenanthroline, a well known receptor for metal ions [62]. A four-channel detection setup combining UV/Vis photometry and fluorimetry, CV and ECL was employed to analyze both the optical and electrochemical changes induced by metal-ion complexation of the QD-crown ether species. CdTe and CdSe QDs of different sizes were coated with the crown ether ligand and titrated with barium cations. A decrease in the QD luminescence intensity upon titration with Ba^{2+} in acetonitrile was observed in all cases, except for the largest (>3 nm) CdTe nanocrystals. CV measurements revealed that the crown ether energy levels complexed with Ba^{2+} ions are shifted below the nanocrystal valence band only for larger CdTe suppressing thus the photo-induced electron transfer mediated by the Ba^{2+} ions justifying the luminescence enhancement for that particular sample. The work constitutes indeed a platform for the construction of quantum dot-based metal ions sensors.

3.7 Applications of Electroactive Quantum Dots

Quantum dots, originally introduced as luminescent labels [2, 7, 8], have more recently emerged as integrated components of assays, chemo- and biosensors, probes, optoelectronic and nanoscale devices [10, 13, 17–19, 73]. Although the majority of these systems rely on optical transduction, some examples based on electrochemical input/output signals have been also reported. These studies are illustrated in the next Sections.

3.7.1 Electrochemical QD Sensors

Examples of quantum dot-based sensors that fully operate in an electrochemical mode are rare in the literature, most likely because of the difficulties associated with electrochemical experiments on semiconductor nanocrystals, as discussed in the first part of this chapter. Thus, it is not surprising that QD-based electrochemical sensors rely on the nanocrystal moiety as an integrated component capable to enhance the electrochemical signal, rather than exploiting the electrochemical response of the quantum dot itself for the sensing function. As it will be illustrated below, these sensors are usually designed to target molecules exhibiting a biological activity.

An electrochemical sensor for the detection of neutravidin, based on core CdTe quantum dots was described (Fig. 3.17) [74]. The CdTe QDs were synthesized in water, and further processed in order to introduce terminal carboxylic groups on the surface, which ultimately offer an anchoring point for the neutravidin conjugation. The as-prepared QDs were activated with 1-ethyl-3-(3-

dimethylaminopropyl) carbodiimide and *N*-hydrohysuccinimide, and finally mixed with a solution of neutravidin to achieve conjugation. A gold electrode was modified with a self-assembled monolayer of cysteamine and subsequently biotinylated. The QD–neutravidin conjugates were captured on top of the modified electrode owing to the ability of biotin to recognize neutravidin. The formation of the sensor assembly was monitored by impedance experiments and atomic force microscopy measurements. Finally, the captured QDs were dissolved in an acid solution to trigger the release of Cd^{2+} ions, whose amount was determined by means of anodic stripping voltammetry while they are electrodeposited on a bismuth film. The intensity of the stripping peak was found to be dependent on the concentration of the QD–neutravidin conjugates and on the extent of their binding onto the electrode surface. In the present system, the QDs play the crucial role of enhancing the current signal, which was not observed in blank experiments (in absence of QDs). The selectivity of the sensor was demonstrated by competing assays with other proteins (horseradish peroxidase, hemoglobin, and myoglobin), which did not lead to a change in the current intensity as observed for the neutravidin. The specificity of the sensor relies on the highly specific and strong interactions that occur between neutravidin and biotin (this receptor–substrate pair has a stability constant as high as 10^{15} M^{-1}). Moreover, the sensor response was proved to be linear between 0.5 and 100 ng/L with a detection limit down to 0.5 nM.

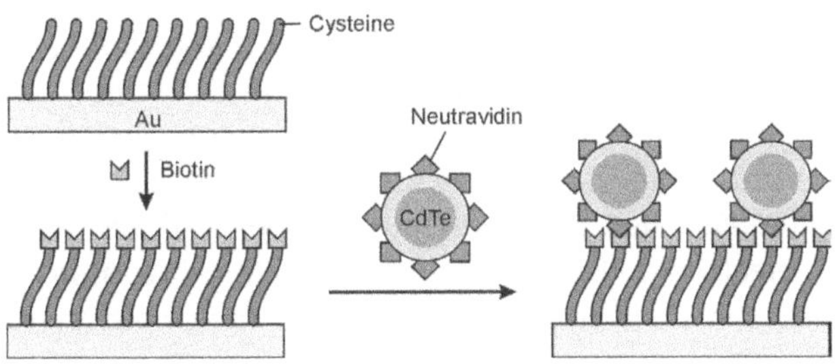

Figure 3.17 The assembly of an electrochemical sensor based on CdTe QDs for the detection of neutravidin [74]. See also Colour Insert.

A very similar approach was used to build an electrochemical immunoassay, based on core–shell CdSe–ZnS quantum dots, for the detection of bovine serum albumin (BSA) [75]. Also in this case, the sensor was found to operate linearly in the range of 0.5–500 ng/mL of phosphorylated BSA, with a detection limit of 0.5 ng/mL. A comparison with the commonly employed method ELISA (enzyme-linked immunosorbent assay) shows that the QD-based sensor has a lower detection limit, thereby demonstrating the applicative potential of these systems for immunologic assays.

Very recently, a light-controlled electrochemical QD sensor was reported [76]. The work relies on the concept of light-addressable potentiometric sensor (LAPS), which is constituted of a semiconductor coupled with an insulating layer. Because of the presence of the latter, a device of this type shows a very poor conductivity as no direct current can flow. The AC conductivity can, however, be increased by illuminating the device with a modulated light, thereby generating electron–hole pairs. The irradiated surface area of the device defines the active spot where an AC photocurrent can flow. The use of a point-shaped irradiation source affords a high spatial resolution (a few microns), despite the fact that the photocurrent flows also in the close periphery of the illuminated area, where electron–hole pairs are generated as well.

Taking rid of the insulating layer and replacing the semiconductor with a QD monolayer, the authors were able to build a device that generates DC photocurrent, in contrast with classical LAPS models that produce AC photocurrents [77]. Particularly, quantum dots of different size and composition, coated with a dithiol ligand, were immobilized on a gold electrode by incubation or spin-coating. The generation of a photocurrent was achieved locally by applying a potential to the electrode and irradiating it with a focused light spot obtained from a common lamp and lenses. The photocurrent measurements were carried out with a potentiostat using a gold film as the working electrode (where the thiolated QDs had been previously deposited) and an Ag/AgCl reference electrode. The effect of the QD material (composition and shape) and the surface immobilization on the photocurrent drift, noise, and influence of the illumination photocurrent were tested in order to optimize the device preparation. Indeed, the authors found a strong dependence of the sensor performances on the manner the quantum dots are attached to the electrode. The layer-by-layer adsorption of a polyelectrolyte film on top of the QDs appeared a convenient solution for technological applications; such an option, however, may impose limitations to the use of the device for analyte detection. The ability of the QDs to exchange electrons with external species was demonstrated by putting the QD-coated gold electrode in contact with solutions of the redox protein cytochrome C and superoxide ions. Moreover, the authors were able to demonstrate that quantum dots can further improve the lateral resolution compared with classic LAPS models. This is because the semiconductor layer is no longer continuous, but composed of grains (i.e., the QDs) that are electronically decoupled, thus preventing any current flow outside the illuminated spot. This preliminary work has shown that functionalized QDs can be employed as light-controlled electrical switches for spatially resolved measurements. An interesting future development of this strategy is its extension for the construction of devices that can sense real-world analytes.

Lin and coworkers recently reported an amperometric biosensor for methyl parathion (MP) [78]. The sensor was built by functionalizing a glassy carbon electrode (GCE) with a multi-walled carbon nanotube (MWCNT) decorated with gold nanoparticles (NPs). The modified GCE was further engineered by the covalent attachment of thiol-coated CdTe quantum dots on which the methyl parathion degrading enzyme (MPDE) was immobilized. The MWCNT/gold NPs element was introduced both to increase the surface area and to catalyze the enzymatic reaction, thus providing a synergetic effect on the signal amplification. The modified GCE was employed as the working electrode in a three-electrode electrochemical cell setup, together with a calomel reference electrode and a platinum wire as the counter electrode. The voltammograms of MP exhibit an anodic peak attributed to the oxidation of *p*-nitrophenol, a product of the MP hydrolysis catalyzed by MPDE. The anodic peak increases on increasing the MP concentration. The sensor shows a detection limit as low as 1.0 ng/mL and a high selectivity for the detection of organophosphates containing the P–S bond. An important feature of the sensor is the possibility to be reused, making it suitable in principle for continuous monitoring.

3.7.2 Electrochemiluminescent QD Sensors

The construction of ECL chemo- and biosensors is a subject of intensive research in the area of quantum dots. In such sensors the QD is either conjugated to other components and hence it constitutes an integrated part of the systems, or it is simply dissolved in the same medium of the analyte or of the other chemicals necessary to develop the ECL signal (e.g., the coreactant).

The first application of the electrogenerated chemiluminescence arising from quantum dots was reported in 2004 by Zou and Ju [55]. Bare CdSe nanocrystals were deposited on top of a paraffin-impregnated graphite electrode (PIGE). The as-prepared sensor was employed to detect hydrogen peroxide in deaerated conditions down to the concentration of 0.1 μM. A linear relationship (Fig. 3.18) between the ECL signal and the hydrogen peroxide concentration was found in a fair range of the analyte concentration (2.5×10^{-7} – 6×10^{-5} M). The operational mechanism of the sensor relies on the ability of the analyte (hydrogen peroxide) to act as a coreactant for the ECL.

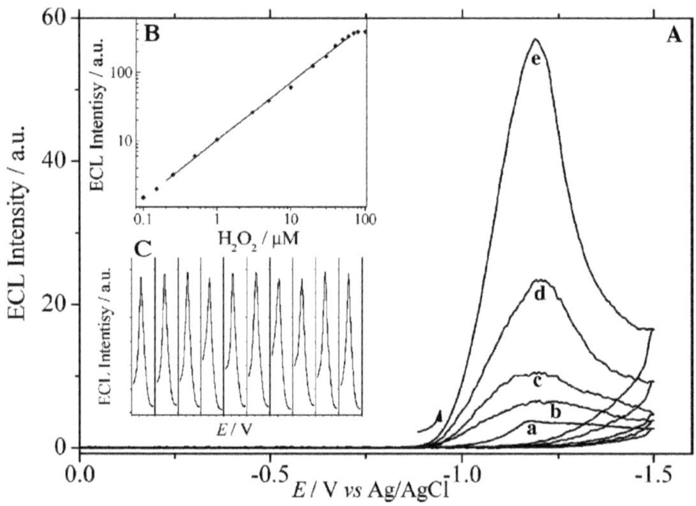

Figure 3.18 (A) Effect of hydrogen peroxide concentration on the ECL intensity generated by a paraffin-impregnated graphite electrode (PIGE) modified with a thin film of CdSe nanocrystals, in the presence of (a) 0, (b) 0.5, (c) 1, (d) 3, and (e) 10 μM H_2O_2. Conditions: 0.1 M PBS, pH 9.3, 0.1 M KNO_3, N_2 saturation for 25 min, 40 mV s^{-1}. (B) Calibration curve for H_2O_2 detection; (C) reproducibility of ECL response of the modified electrode in the presence of 10 μM H_2O_2 under the same conditions as (A). From ref. 55.

The same authors extended this principle to glucose sensing [79]. CdSe QDs capped with thioglycolic acid (TGA) were bioconjugated with glucose oxidase (GOx) by successive drop casting of appropriate amounts of each solution onto a PIGE. The GOx/TGA–CdSe architectures exhibited a lower resistance than that of the film composed of GOx–TGA conjugates, thus enabling a faster electron transfer between the nanocrystals and the electrode. This is indeed the first example of a biosensor based on the intrinsic cathodic ECL signal emerging from colloidal semiconductor nanocrystals. Both oxygen and hydrogen peroxide were found to work as ECL coreactants, but oxygen resulted to be more efficient because it can capture more electrons from the reduced QDs. Oxygen-saturated solutions exhibited two distinct cathodic ECL peaks, attributed respectively to the free and conjugated quantum dots. On addition of glucose to the GOx/TGA–CdSe solution the ECL signal decreases, an observation that can be rationalized on the basis of the glucose cycle. Owing to the catalytic action of GOx, the oxygen dissolved in solution is transformed into an equimolar amount of hydrogen peroxide, which is a less effective ECL coreactant compared with oxygen. Hence, the ECL decrease is due to the conversion of oxygen into hydrogen peroxide, thus providing a platform for glucose sensing with the described nanohybrid. The concentration of the nanoconjugate, the size of the QDs, and the pH of the solution (which affects the electrostatic interactions between the anion-capped QDs) were all found to be critical parameters for the generation of the electrochemiluminescence signal. The ECL response to glucose was found to be linear in a wide range of concentrations (2.5×10^{-5} – 3×10^{-3} M) and with a detection limit (4.0 μM) lower than other ECL-based glucose sensors [80].

The above described concept was employed for the detection of other molecules of biological interest (glutathione (GSH) and L-cysteine), exploiting hydrogen peroxide as a coreactant for the cathodic ECL signal. A film obtained by the deposition of TGA–CdSe QDs onto PIGE, placed in an aqueous solution of hydrogen peroxide, was used for the detection of both scavengers and generators of the hydroxyl radical [81]. Initially, the CdSe QDs are reduced to form negatively charged particles which are thus able to convert hydrogen peroxide into hydroxyl radicals. The formation of these radicals, whose presence was confirmed by electron paramagnetic resonance (EPR) spectroscopy, strongly affects the ECL mechanism. The ECL emission of TGA–CdSe QDs resulted to be quenched by annihilation with the hydroxyl radicals, thereby offering a switchable signal useful for sensing thiol-containing biological compounds, which are efficiently oxidized by hydroxyl radical. The described sensor can detect GSH and L-cysteine with a linear behavior in a respectable concentration range (2.0–60 μM for GSH and 2.0–50 μM for L-cysteine). The essay was found to be unaffected by the presence of other amino acids, whereas dopamine interfered with the analysis, although to a small extent, because of its ability to react with hydroxyl radicals.

Weak points of the early examples of electrochemiluminescent quantum dot sensors are the poor stability and a lower ECL signal intensity compared with other luminophores. Therefore, efforts have been undertaken on the development of more robust systems capable of producing an intense ECL signal. Carbon nanotubes (CNTs) were found to enhance the electrochemiluminescence intensity of CdS QDs as well as decrease its starting potential value [82]. CdS core QDs coated with thioglycolic acid (acting here as a protective agent) were combined with carbon nanotubes and deposited on an electrode. When negative potentials were applied to such a modified electrode in the presence of hydrogen peroxide, a strong ECL signal was produced. Such a signal resulted to be 5 times higher than that arising from a film of the sole CdS QDs. The role of carbon nanotubes was evaluated by means of cyclic voltammetry. The results show that the onset reduction potential of the CdS QDs is shifted to less negative values and the cathodic peak current is enhanced upon incorporation of the nanocrystals into the composite thin film. Presumably, the carbon nanotubes can mediate the electron-transfer process, thereby reducing the potential barrier for electron injection into the CdS nanocrystals. Moreover, owing to the porous structure of the composite layer, hydrogen peroxide molecules can diffuse into the film, thus enabling ECL generation both at the surface and in the bulk of the film itself. The composite layer did not show degradation after 30 scan cycles in the reduction window.

Au/CdS nanostructured films also showed an enhanced ECL signal, and were employed for hydrogen peroxide sensing [83]. The composite film was prepared simply by mixing amounts of gold nanoparticles and CdS quantum dots in water, and subjecting the mixture to ultrasonication. The as-prepared film generates a cathodic ECL signal in hydrogen peroxide solution. The proposed operation mechanism is analogous to that discussed for the previous example: the electrochemically reduced CdS quantum dots react with hydrogen peroxide, thereby generating excitons in the QDs, which subsequently recombine with the emission of a photon. The ECL intensity of the nanocomposite film was found to be four times higher than that observed for the bare CdS QD film. In this case the ECL signal enhancement was attributed to a stabilization exerted by the Au nanoparticles on the reduced CdS QDs.

Yang and coworkers showed recently the ability of silver(I) ions to enhance the electrogenerated chemiluminescence signal of CdTe quantum dots [84]. The CdTe QDs were synthesized in water and immobilized onto a glassy carbon electrode via a layer-by-layer assembly together with poly(diallyldimethylammonium chloride) (PDADMAC), which offers anchoring points for the functionalization, and 2-(dibutylamino)-ethanol, acting as the ECL coreactant. After the addition of Ag^+ ions a fourfold ECL enhancement was observed, both in the presence and in the absence of the coreactant.

Once it became clear that a large variety of species could be used to enhance the electrochemiluminescence signal of QDs, many ECL chemo- and biosensors have been developed. The QD–CNT conjugates have been used to build both immunosensors [85] and a biosensor for acetylcholine and choline [86]. The gold nanoparticle/quantum dot composites have been employed to devise immunoassays for the detection of human prealbumin [87], C-reactive protein [88] and a low-density lipoprotein [89]. As for the previously described cases, the detection was based on the analyte-dependent depletion of the ECL signal. The sensors show a linear behavior over a wide concentration range, and low detection limits (down to a few pg/mL).

The examples reported so far exploit cathodic ECL and imply the use of a coreactant to produce the signal. Investigations dealing with anodic ECL [66] and electrochemiluminescence generated by direct annihilation of QDs [65, 90] are also available.

3.7.3 QD-Based Electrochemical Signal Transducers

A 2D assembly of quantum dots was employed to acheive electrochemical rectification [91]. A molecular blocking layer was constructed by self-assembling 11-mercaptoundecanoic acid (MUA) and an ultrathin PDADMAC film directly on the electrode. CdTe QDs coated with 3-mercaptopropionic acid (MPA) were electrostatically adsorbed on the polymer film to obtain the 2D assembly. The QDs adsorbed onto the thiol-modified gold electrodes, in contact with a solution of the redox couple $[Fe(CN)_6]^{3-/4-}$, show a nonsymmetric enhancement of the oxidation and reduction currents, with the cathodic signal undergoing a larger increase than the anodic one (Fig. 3.19). The authors found a dependence of the electron-transfer rate on the direction of the electron flow, a behavior that is usually referred to as rectification. The extent of the electrochemical rectification is related to the size of the QDs, enabling an interpretation based on the energy position of the valence and conduction band edges of the nanocrystals relatively to the redox potential of the electroactive species in solution. Considering that the energy of the conduction band is considerably higher than the level of the redox couple, the mediated electron transfer occurs predominantly via the valence band. In fact, when the nanocrystal size is decreased, the energy of the valence band is shifted well below the level of the redox couple, thus requiring an overpotential to generate the holes necessary for the oxidation process. This interpretation is in agreement with the experimental observation that the anodic peak becomes more intense upon increasing the average diameter of the QDs.

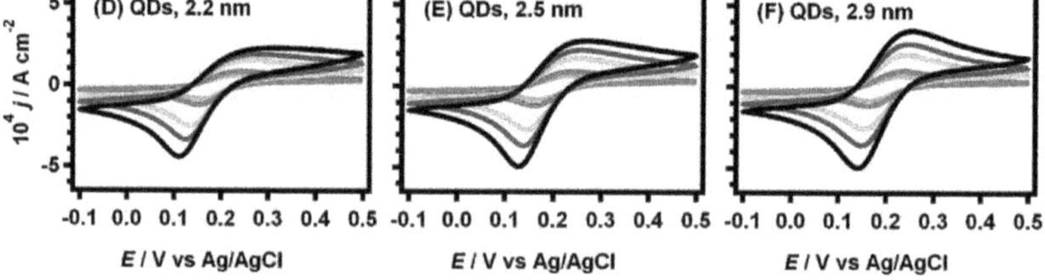

Figure 3.19 Cyclic voltammograms in the presence of 1.0 mM $K_4[Fe(CN)_6]$ and $K_3[Fe(CN)_6]$ for a clean Au electrode (A) and after modification with MUA (B), MUA-PDADMAC (C) (insets representing 14 × magnification), and MUA-PDADMAC-QDs of diameters 2.2 (D), 2.5 (E), and 2.9 nm (F). The voltammograms were recorded at 5, 10, 50, 100, and 200 mV s^{-1}. See the text for details. From ref. 91. See also Colour Insert.

Electrochemically induced luminescence changes were studied in air-equilibrated aqueous solution by Knoll and coworkers [92]. The work builds upon previous studies which show that the photoluminescence intensity of quantum dots deposited on gold, platinum and indium tin oxide (ITO) could be controlled electrochemically [44, 57, 93]. Quantum dots made with a CdZnSe alloy and capped with mercaptoundecanoic acid (MUA) were assembled on gold films. The binding of the MUA–QDs on the gold surface was verified with plasmon spectroscopy. A three-electrode cell with the gold film as the reference electrode, a platinum wire as the counter electrode, and an Ag/AgCl reference electrode, was constructed. The surface plasmon enhanced fluorescence spectra (SPFS) of the quantum dots deposited on the electrode was monitored as a function of the applied electrical potential. The luminescence signal increases by one order of magnitude when the QDs are deposited on the Au-SAM electrode in comparison with the free electrode. Electrochemical measurements under illumination revealed the occurrence of an anodic current. The origin of such a current was attributed to the transfer of optically excited electrons in the quantum dot to the gold; the hole left behind in the quantum dot is filled by electron donation from sulphite ions in the solution. Further experiments indicate that the QD SPFS is highly dependent on the applied potential; specifically, the luminescence signal is suppressed upon application of negative potentials. Such a luminescence switching is reversible, as a full recover was observed when the cell was brought back to 0 V (versus Ag/AgCl) or to open circuit. The possibility of controlling the QD luminescence by electrochemical stimulation in an aqueous environment might be useful for new biosensing applications.

3.7.4 Other Systems

Very recently, QD–peptide–dopamine conjugates were constructed and investigated (Fig. 3.19) [94]. Dopamine, a well known neurotransmitter, is characterized by an interesting redox behavior which includes (i) a coupled electron-proton transfer process, whose formal reduction potential depends linearly on pH, and (ii) oxidation of hydroquinone to the quinone form by O_2 at basic pH. CdSe–ZnS QDs capped with PEG-modified dihydrolipoic acid ligands (DHLA–PEG) were self-assembled with a dopamine-functionalized peptide, and their behavior in aqueous solution as a function of pH was investigated by electrochemistry and spectroscopy. At low pH values, when the dopamine is in the hydroquinone form, no quenching of the QD emission takes place. As the pH is increased, the formal reduction potential of dopamine becomes less negative; moreover, oxidation of the hydroquinone to the quinone form by dioxygen takes place. The quinone is a strong electron acceptor and can effectively quench the QD luminescence by electron transfer. As a result, the QD–peptide–dopamine conjugates exhibit a luminescence output which depends linearly on pH. These systems were also used to sense cytoplasmic pH in cells undergoing drug-induced alkalosis [94].

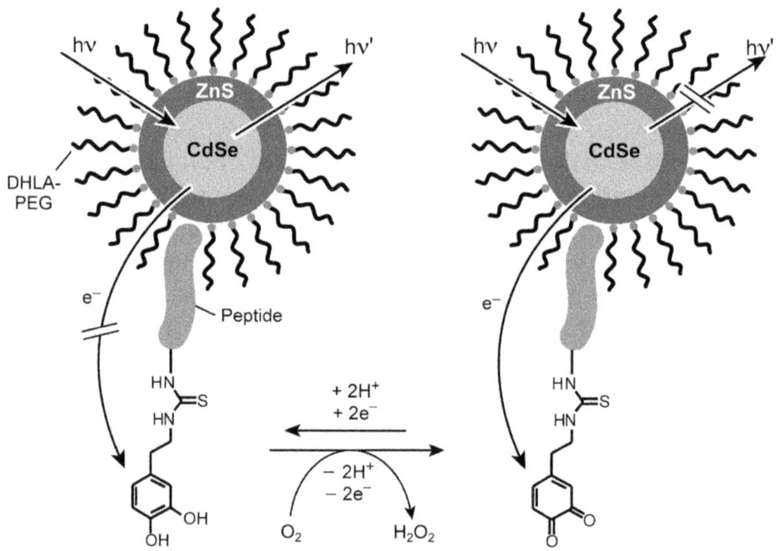

Figure 3.20 CdSe–ZnS QDs modified with a peptide-dopamine species are capable of exploiting the pH-dependent redox properties of dopamine to sense pH in aqueous solutions and in biological environments [94]. See also Colour Insert.

A very interesting and elegant example of an electrochemically controllable QD-based nanohybrid was recently reported [95]. A boron nitride nanotube coated with a thin layer of gold, acting as both an electrode and a nanoneedle, was attached to a self-assembled monolayer composed of streptavidin-conjugated CdSe–ZnS core–shell quantum dots (Fig. 3.21a). Controlled release of the dots was achieved by means of the electrochemical desorption of

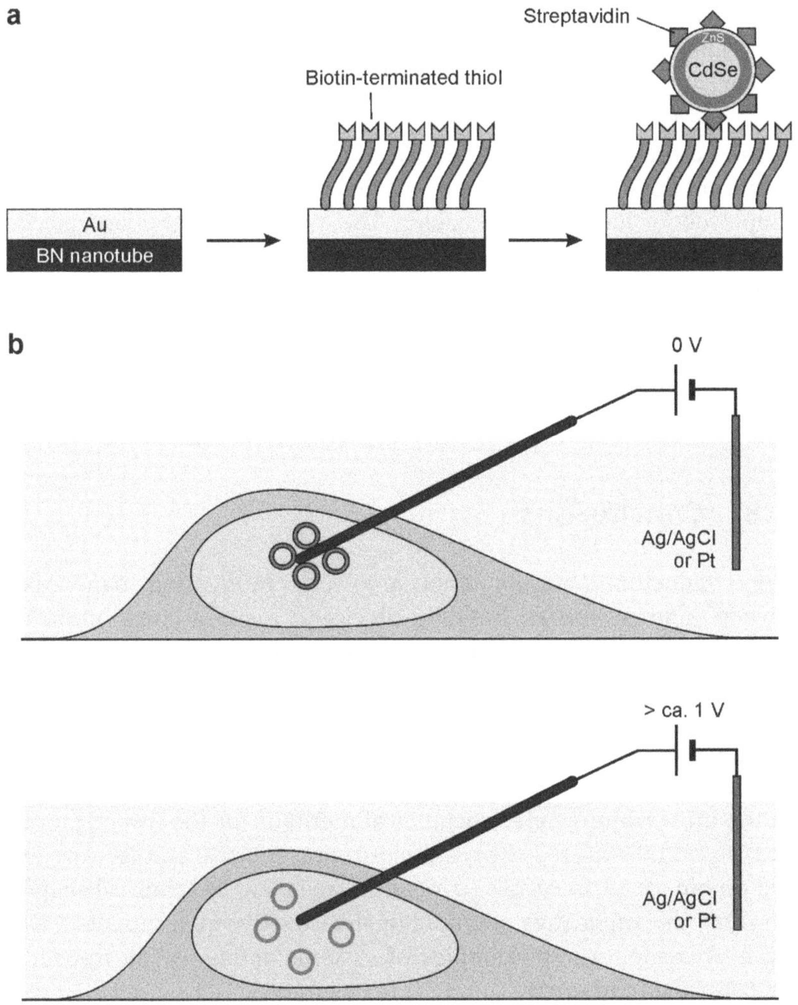

Figure 3.21 (a) Procedure for surface functionalization of a gold-covered nanoneedle and conjugation of streptavidin-coated CdSe–ZnS core–shell QDs (Invitrogen QDot 655). (b) Schematic representation of the electrochemically controlled release of the QDs inside a cell [95]. See also Colour Insert.

the quantum dot monolayer from the gold coating of the nanotube. The release was also demonstrated by delivering the QDs into the nucleus of living HeLa cells. After the nanoneedle reached the target area, an electrical potential was applied for a few seconds in order to release the QD monolayer (Fig. 3.21b). The release can be accurately controlled by changing the magnitude and the duration of the applied electrical potential. The nanoneedle is fully regenerable because once the QD layer has been fully detached from the gold surface, it can be reused for a new deposition and then a new release. The quantum dots, once released in the desired cell compartment, were employed for fluorescence imaging. The method appears particularly convenient in the light of previous work which demonstrated that QDs of similar size could not penetrate into the cell nucleus [96]. The system, in principle, can be modified to operate in different settings by changing the chemical nature of the load; it appears particularly promising for applications in which an external control on the load release is required (e.g., drug delivery).

3.8 Conclusions

Semiconductor quantum dots are very interesting nanoscale objects, and promising building blocks to make multicomponent systems that perform functions taking advantage of electron-transfer processes. Hence, the study of the electrochemical behavior of quantum dots is important not only for basic science reasons but also to foster the development of QD-based systems amenable of practical applications. In this chapter, we have discussed the use of the most common electrochemical methods for the investigation of QDs, and illustrated the results of representative studies carried out on binary nanocrystals (core and core–shell) of spherical shape. To show the applicative potential of electroactive quantum dots, we have presented a few examples of systems acting as (bio)sensors and signal transducers.

The ability of electrochemistry to probe the surface of QDs is of high utility, particularly if electrochemical experiments are complemented with spectroscopic ones. However, the ill-defined redox behavior often observed for nanocrystals, with the consequent difficulty to probe their redox levels, complicates substantially the design of QD-based devices and materials exhibiting predetermined functionalities. Indeed, further fundamental investigations are needed to see if and how the vast amount of knowledge accumulated in the modular construction of molecular devices can be transferred to semiconductor nanocrystals.

Acknowledgments

Financial support from the European Union (FP7 HYSENS project, grant no. 263091), Ministero dell'Istruzione, Università e Ricerca (PRIN), Ministero degli Affari Esteri (Direzione Generale per la Promozione del Sistema Paese), and Università di Bologna is gratefully acknowledged.

References

1. Bawendi, M. G., Steigerwald, M. L., Brus, L. E. *Annu. Rev. Phys. Chem.*, 1990, **41**, 477–496.
2. Alivisatos, A. P. *Science*, 1996, **271**, 933–937.
3. Yoffe, A. D. *Adv. Phys.*, 2001, **50**, 1–208.
4. Efros, A. L., Rosen, M. *Annu. Rev. Mater. Sci.*, 2000, **30**, 475–521.
5. Burda, C., Chen, X. B., Narayana, R., El-Sayed, M. A. *Chem. Rev.*, 2005, **105**, 1025–1102.
6. *Semiconductor Nanocrystal Quantum Dots*, Rogach, A. L., Ed., Springer: Wien, Austria, 2008.
7. Michalet, X., Pinaud, F. F., Bentolila, L. A., Tsay, J. M., Doose, S., Li, J. J., Sundaresan, G., Wu, A. M., Gambhir, S. S., Weiss, S. *Science*, 2005, **307**, 538–544.
8. Resch-Genger, U., Grabolle, M., Cavaliere-Jaricot, S., Nitschke, R., Nann. T. *Nat. Meth.*, 2008, **5**, 763–775.
9. Green, M. *J. Mater. Chem.*, 2010, **20**, 5797–5809.
10. (a) Callan, J. F. , De Silva, A. P., Mulrooney, R. C., McCaughan, B. *J. Incl. Phenom. Macroc. Chem.*, 2007, **58**, 257–262; (b) Raymo, F. M., Yildiz, I. *Phys. Chem. Chem. Phys.*, 2007, **9**, 2036–2043. (c) Medintz, I. L., Mattoussi, H. *Phys. Chem. Chem. Phys.*, 2009, **11**, 17–45.
11. de Silva, A. P., Gunaratne, H. Q. N., Gunnlaugsson, T., Huxley, A. J. M., McCoy, C. P., Rademacher, J. T., Rice, T. E. *Chem. Rev.*, 1997, **97**, 1515–1566.
12. Balzani, V., Credi, A., Venturi, M. *Molecular Devices and Machines – Concepts and Perspectives for the Nanoworld*, Wiley-VCH: Weinheim, 2008.
13. (a) Yildiz, I., Tomasulo, M., Raymo, F. M. *J. Mater. Chem.*, 2008, **18**, 5577–5584; (b) Amelia, M., Avellini, T., Semeraro, M., Silvi, S., Credi, A. *Curr. Phys. Chem.*, in press.

14. *Electron transfer in Chemistry*, V. Balzani, Ed., Wiley-VCH: Weinheim, 2001.
15. Kaifer, A. E., Gomez-Kaifer, M. *Supramolecular Electrochemistry*, Wiley-VCH: Weinheim, 1999.
16. *Electrochemistry of Functional Supramolecular Systems*, Ceroni, P., Credi, A., Venturi M., Eds., John Wiley: New York, 2010.
17. Somers, R. C., Bawendi, M. G., Nocera, D. G. *Chem. Soc. Rev.*, 2007, **36**, 579–591.
18. Yildiz, I., Deniz, E., Raymo, F. M., *Chem. Soc. Rev.*, 2009, **38**, 1859–1867.
19. Talapin, D. V., Lee, J. S., Kovalenko, M. V., Shevchenko, E. V. *Chem. Rev.*, 2010, **110**, 389–458.
20. Smith, A. M., Nie, S. *Acc. Chem. Res.*, 2010, **43**, 190–200.
21. Bard, A. J., Ding, Z., Myung, N., *Struct. Bond.*, 2005, **118**, 1–57.
22. Norris, D. J., Efros, A. L., Rosen, M., Bawendi, M. G. *Phys. Rev. B*, 1996, **53**, 16347–16354.
23. Bard, A. J., Faulkner, L. R. *Electrochemical Methods – Fundamentals and Applications, 2nd Edition*, Wiley: New York, 2000.
24. Franceschetti, A., Zunger, A. *Phys. Rev. B*, 2000, **62**, 2614–2623.
25. Hole addition corresponds to the extraction of one electron from the nanocrystal.
26. Kuçur, E., Bücking, W., Giernoth, R., Nann, T. *J. Phys. Chem. B*, 2005, **109**, 20355–20360.
27. Guyot-Sionnest, P., Wang, C. *J. Phys. Chem. B*, 2003, **107**, 7355–7359.
28. Inamdar, S. N., Ingole, P. P., Haram, S. K. *Chem. Phys. Chem.*, 2008, **9**, 2574–2579.
29. *Electrogenerated Chemiluminescence* (Ed.: A. J. Bard), Marcel Dekker, New York, 2004.
30. Yu, W. W, Qu, L. H., Guo, W. Z., Peng, X. G. *Chem. Mater.*, 2003, **15**, 2854–2860.
31. Wang, X. Y, Qu, L. H, Zhang, J. Y, Peng, X. G, Xiao, M. *Nano Lett.*, 2003, **3**, 1103–1106.
32. Yu, W. W, Peng, X. G. *Angew. Chem. Int. Ed.*, 2002, **41**, 2368–2371.
33. Qu, L. H., Peng, X. G *J. Am. Chem. Soc.*, 2002, **124**, 2049–2055.
34. Peng, Z. A., Peng, X. *J. Am. Chem. Soc.*, 2001, **123**, 1389–1395.
35. Peng, X. G., Manna, L., Yang, W. D., Wickham J., Scher, E., Kadavanich, A., Alivisatos, A. P. *Nature*, 2000, **404**, 59–61.

36. Kuçur, E., Riegler, J., Urban, G. A., Nann, T. *J. Chem Phys.*, 2003, **119**, 2333–2337.

37. Querner, C., Reiss, P., Sadki, S., Zagorskab, M., Pron, A. *Phys. Chem. Chem. Phys.*, 2005, **7**, 3204–3209.

38. Amelia, M., Avellini, T., Monaco, S., Impellizzeri, S., Yildiz, I., Raymo, F. M., Credi. A. *Pure Appl. Chem.*, 2011, **83**, 1–8.

39. Amelia, M., Impellizzeri, S., Monaco, S., Yildiz, I., Silvi, S., Raymo, F. M., Credi, A. submitted.

40. Donega, C. D., Liljeroth, P., Vanmaekelbergh, D. *Small*, 2005, **1**, 1153–1162.

41. Rogach, A. L., Talapin, D. V., Shevchenko, E. V., Kornowski, A., Haase, M., Weller, H. *Adv. Funct. Mater.*, 2002, **12**, 653–644.

42. Wang, C., Shim, S., Guyot-Sionnest, P. *Appl. Phys. Lett.*, 2002, **80**, 4–6.

43. Wehrenberg, B. L., Guyot-Sionnest, P. *J. Am. Chem. Soc.*, 2003, **125**, 7806–7807.

44. Houtepen, A. J., Vanmaekelbergh, D. *J. Phys. Chem. B*, 2005, **109**, 19634–19642.

45. Matylitsky, V., Tworak, L., Breus, V., Basché, T., Wachtveitl J. *J. Am. Chem. Soc.*, 2009, **131**, 2424–2425.

46. Gooding, A. K., Gómez, D. E., Mulvaney, P. *ACS Nano*, 2008, **2**, 669–676.

47. Dabbousi, B. O., Rodriguez-Viejo, J., Mikulec, F. V., Heine, J. R., Mattoussi, H., Ober, R., Jensen, K. F., Bawendi, M. G. *J. Phys. Chem. B*, 1997, **101**, 9463–9475.

48. Impellizzeri, S., Monaco, S., Yildiz, I., Amelia, M., Credi, A., Raymo, F. M. *J. Phys. Chem. C*, 2010, **114**, 7007–7013.

49. Reiss, P., Protière, M., Li, L. *Small*, 2009, **5**, 154–168.

50. Myung, N., Ding, Z., Bard, A. J. *Nano Lett.*, 2002, **2**, 1315–1319.

51. Zhou, J., Zhu, J., Brzezinski, J., Ding, Z., *Can. J. Chem.*, 2009, **87**, 386–391.

52. Myung, N., Bae, Y., Bard, A.J. *Nano Lett.*, 2003, **3**, 1053–1055.

53. Poznyak, S. K., Talapin, D. V., Shevchenko, E. V., Weller, H. *Nano Lett.*, 2004, **4**, 693–698.

54. Richter, M. M. *Chem. Rev.*, 2004, **104**, 3003–3036.

55. Zou, G., Ju, H. *Anal. Chem.*, 2004, **76**, 6871–6876.

56. Wang, V., Shim, M., Guyot-Sionnest, P. *Science*, 2001, **291**, 2390–2392.

57. Cordero, S. R., Carson, P. J., Estabrook, R. A., Strouse, G. F., Buratto, S. K. *J. Phys. Chem. B*, 2000, **104**, 12137–12142.

58. Bae, Y., Myung, N., Bard, A. J. *Nano Lett.*, 2004, **4**, 1153–1161.

59. Poznyak, S. K., Osipovich, N. P., Shavel, A., Talapin, D. V., Gao, M., Eychmüller, A., Gaponik, N. *J. Phys. Chem. B*, 2005, **109**, 1094–1100.

60. Borchert, H., Talapin, D. V., Gaponik, N., McGinley, C., Adam, S., Lobo, A., Möller, T., Weller, H. *J. Phys. Chem. B*, 2003, **107**, 9662–9668.

61. Haram, S. K., Quinn, B. M., Bard, A. J. *J. Am. Chem. Soc.*, 2001, **123**, 8860–8861.

62. Völker, J., Zhou, X., Ma, X., Flessau, S., Lin, H., Schmittel, M., Mews, A. *Angew. Chem. Int. Ed.*, 2010, **49**, 6865–6868.

63. (a) Talapin, D. V., Rogach, A. L., Kornowski, A., Haase, M., Weller, H. *Nano Lett.*, 2001, **1**, 207–211; (b) Wuister, S. F., Donega, C. D., Meijerink, A. *J. Phys. Chem. B*, 2004, **108**, 17393–17397; (c) Wang, J.-H., Wang, H.-Q., Zhang, H.-L., Li, X.-Q., Hua, X.-F., Huang, Z.-L., Zhao, Y.-D. *Colloids and Surfaces A: Physicochem. Eng. Aspects*, 2007, **305**, 48–53; (d) Li, L., Qian, H., Fang, N., Ren, J. *J. Luminescence*, 2006, **116**, 59–66.

64. Han, H., Sheng, Z., Liang, J. *Analytica Chimica Acta*, 2007, **596**, 73–78.

65. Hua, L., Han, H., Zhang, X., *Talanta*, 2009, **77**, 1654–1659.

66. Liu, X., Jiang, H., Lei, J., Ju, H. *Anal. Chem.*, 2007, **79**, 8055–8060.

67. Liu, X., Ju, H. *Anal. Chem.*, 2008, **80**, 5377–5382.

68. Zhang, L., Zou, X., Ying, E., Dong, S. *J. Phys. Chem. C*, 2008, **112**, 4451–4454.

69. Jin, L., Shang, L., Zhai, J., Li, J., Dong, S. *J. Phys. Chem. C*, 2010, **114**, 803–807.

70. Torimoto, T., Maeda, K., Maenaka, J., Yoneyama, H. *J. Phys. Chem.*, 1994, **98**, 13658–13664.

71. (a) Dorokhin, D., Tomezak, N., Han, M. Y., Reinhoudt, D. N., Velders, A. H., Vancso, G. J. *ACS Nano*, 2009, **3**, 661–667; (b) Dorokhin, D., Tomezak, N., Velders, A. H., Reinhoudt, D. N., Vancso, G. J. *J. Phys. Chem. C*, 2009, **113**, 18676–18680.

72. Dorokhin, D., Tomezak, N., Reinhoudt, D. N., Velders, A. H., Vancso, G. J. *Nanotechnology*, 2010, **21**, 285703/1–285703/6.

73. (a) Wang, X. J., Ruedas-Rama, M. J., Hall, E. A. H. *Anal. Lett.*, 2007, **40**, 1497–1520; (b) Gill, R., Zayats, M., Willner, I. *Angew. Chem. Int. Ed.*, 2008, **47**, 7602–7625; (c) Alivisatos, A. P., Gu, W., Larabell, C. *Annu. Rev. Biomed. Eng.*, 2005, **7**, 55–76.

74. Dan, D., Ding, J., Tao, Y., Li, H., Chen, X. *Biosens. Bioelectron.*, 2008, **24**, 863–868.

75. Pinwattana, K., Wang, J., Lin, C.-T., Wu, H., Du, D., Lin, Y., Chailapakul, O. *Biosens. Bioelectron.*, 2010, **26**, 1109–1113.

76. Yue, Z., Khalid, W., Zanella, M., Abbassi, A. Z., Pfreundt, A., River Gil, P., Schubert, K., Lisdat, F., Parak, W. J. *Anal. Biochem. Chem.*, 2010, **396**, 1095–1103.

77. (a) George, M., Parak, W. J., Gaub, H. E. *Sens. Actuators B*, 2000, **69**, 266–275; (b) Owicki, J. C., Bousse, L., Hafeman, D., Kirk, G. L., Olson, J. D., Wada, H. G., Parce, J. W. *Annu. Rev. Biophys. Biomol. Struct.*, 1994, **23**, 87–113.

78. Dan, D., Chen, W., Zhang, W., Liu, D., Li, H., Lin, Y. *Biosens. Bioelectron.*, 2010, **25**, 1370–1375.

79. Jiang, H., Ju, H. *Chem. Commun.*, 2007, 404–406.

80. (a) Corgier, B. P., Marquette, C. A., Blum, L. J. *Anal.* Chim. *Acta*, 2005, **538**, 1–7; (b) Marquette, C. A., Degiuli, A., Blum, L. J. *Biosens. Bioelectron.*, 2003, **19**, 433–439; (c) Marquette, C. A., Blum, L. J. *Biosens. Bioelectron.*, 2004, **20**, 197–203; (d) Jameison, F., Sanchez, R. I., Dong, L., Leland, J. K., Yost, D., Martin, M. T. *Anal. Chem.*, 1996, **68**, 1298–1302; (e) Martin, A. F., Nieman, T. A. *Biosens. Bioelectron.*, 1997, **12**, 479–489.

81. Jiang, H., Ju, H. *Anal. Chem.*, 2007, **79**, 6690–6696.

82. Ding, S.-N., Xu, J.-J., Chen, H.-Y. *Chem. Commun.*, 2006, 3631–3633.

83. Shi, C., Shan, Y., Xu, J., Chen, H. *Electrochim. Acta*, 2010, 8268–8272.

84. Li, Y., Sun, M., Yang, F., Yang, X. *Anal. Lett.*, 2010, **43**, 2837–2847.

85. (a) Jie, G. , Liu, P., Wang, L., Zhang, S. *Electrochem. Comm.*, 2010, **12**, 22–26; (b) Jie, G., Zhang, J., Wang, D., Chen, C., Chen, H. Y., Zhu, J. J. *Anal. Chem.*, 2008, **80**, 4033–4039; (c) Jie, G., Li, L., Chen, C., Xuan, J., Zhu, J. J. *Biosens. Bioelectron.*, 2009, **24**, 3352–3358.

86. Wang, X. F., Zhou, Y., Xu, J. J., Chen, H. Y. *Adv. Funct. Mater.*, 2009, **19**, 1444–1450.

87. Jie, G., Huang, H., Sun, X., Zhu, J. J. *Biosens. Bioelectron.*, 2008, **23**, 1896–1899.

88. Wang, S., Harris, E., Shi, J., Chen, A., Parajuli, S., Jing, X., Miao, W. *Phys. Chem. Chem. Phys.*, 2010, **12**, 10073–10080.

89. Jie, G., Liu, B., Pan, H., Zhu, J. J., Chen, H. Y. *Anal. Chem.*, 2007, **79**, 5574–5581.

90. Hua, L., Zhou, J., Han, H. *Electrochim. Acta*, 2010, **55**, 1265–1271.

91. Kissling, G. P., Bünzli, C., Fermín, D. J. *J. Am. Chem. Soc.*, 2010, **132**, 16855–16861.

92. Cameron, P. J., Zhong, X., Knoll, W. *J. Phys. Chem. C*, 2009, **113**, 6003–6008.
93. (a) Wang, C., Wehrenberg, B. L., Woo, C. Y., Guyot-Sionnest, P. *J. Phys. Chem. B*, 2004, **108**, 9027–9031; (b) Shim, M., Wang, C., Guyot-Sionnest, P. *J. Phys. Chem. B*, 2001, **1058**, 2369–2373; (c) Jha, P. P., Guyot-Sionnest, P. *J. Phys. Chem. C*, 2007, **111**, 15440–15445; (d) Myung, N., Bae, Y., Bard, A. J. *Nano Lett.*, 2003, **3**, 747–749.
94. Medintz, I. L., Stewart, M. H., Trammell, S. A., Susumu, K., Delehanty, J. B., Mei, B. C., Melinger, J. S., Blanco-Canosa, J. B., Dawson, P. E., Mattoussi, H. *Nature Mater.*, 2010, **9**, 676–684.
95. Yum, K., Wang, N., Yu, M.-F. *Small*, 2010, **6**, 2109–2113.
96. Chen, F., Gerion, D. *Nano Lett.*, 2004, **4**, 1827–1932.

Chapter 4

Electron Transfer Quenching for Biosensing with Quantum Dots

David E. Benson,*[*] Stacey R. De Haan, Chase E. Hulderman, and Marla D. Swain

Calvin College, Grand Rapids, MI 49546, USA

[*]dbenson@calvin.edu

Sensor design has used electron transfer (ET) increasingly over the past decade. Both chemically and biochemically based sensors (termed chemo- and biosensors) that employ ET as the sensing mode have drawn on decades of electron transfer research in chemistry [1], biology [2], and materials science [3]. This chapter focuses on electron transfer reactions specifically with semiconductor nanoparticles, or quantum dots, that have been employed in biosensing. The primary material is CdSe coated with 2–4 monolayers of ZnS (so-called core–shell nanoparticles) [4, 5]. Due to the wide excitation profile of quantum dots, electron transfer-based sensing methods are not constrained to the emission wavelengths provided by CdSe [6]. As a result, fluorescence emission from 450 nm to 700 nm can be achieved depending on the diameter and material of the water-soluble quantum dot employed. This provides a significant modularity to these systems in addition to their high photostability.

Quantum Dot Sensors: Technology and Commercial Applications
Edited by John F. Callan and Françisco M. Raymo

ISBN 978-981-4316-00-2 (Hardcover), 978-981-4364-61-4 (eBook)
www.panstanford.com

Sensing based on ET with quantum dots provides a combination of modularity and selectivity that is difficult to match with other sensing methods. The ideal method would allow every component to be easily substituted for another component for the desired response (e.g., ligand type, ligand sensitivity, excitation wavelength, emission wavelength). Irreversible quenching methods such as precipitation and decomposition assays can be sensitive but often struggle with selectivity due to the inherent low water solubility of quantum dots [7]. Furthermore, if enzymatic sensing components are introduced these methods often work for one or two analytes [6]. Energy transfer methods provide modularity from the ligand type and sensitivity viewpoint; however, they require a specific pairing of an organic fluorophore with a specific diameter and material of quantum dot [8]. Thus, there are limited fluorophore/quantum dot combinations that can be employed along with the high photobleaching potential of the organic fluorophore.

Using ET quenching as the sensing mode with quantum dots provides modular control of the analyte type and sensitivity. The components in the ET quenching system are the electron donor/acceptor (attached to the biomolecule), the biomolecule that specifically binds a ligand/analyte, and the quantum dot. To date, there are only small changes in redox thermodynamics upon analyte/ligand binding for combinations of electron donors/acceptors and quantum dots. A similar statement can be made for chemosensors [9]. Biosensors, however, provide the significant advantage of modularity [10]. Once the electron donor acceptor has a thiol-reactive functional group (maleimide, iodoacetamide, etc.) various biomolecules can be introduced to the same biomolecule-modified quantum dot to provide a variety of systems. These systems can have a varied reduction potential or surface charge for the electron donor/acceptor. More important, the emission and longest excitation wavelength can be altered to meet the needs of a particular fluorescence detection scheme by changing the type of quantum dot attached to the biomolecule. Finally, the biomolecule can be altered to change analyte specificity (e.g., maltose binding protein to phosphate binding protein) or sensitivity (e.g., amino acid sidechain substitution). The unifying concept is the use of ET quenching to tie all three components together into a functional system.

4.1 Electron Transfer Quenching

Photoinduced electron transfer has been used for sensing and to understand the speed at which electrons propagate through compounds, biomolecules, and materials. In the classic case [11], an excited state is formed in a fluorophore by absorbing a photon of light. The excited state has an electron in a previously unoccupied high-energy molecular orbital leaving an electron vacancy (hole) in the molecular orbital that was originally filled. This makes the fluorophore excited state both a good reductant (electron transfer from a high-energy molecular orbital to an electron acceptor) and a good oxidant (electron transfer from an electron donor to a hole in the originally filled molecular orbital). These charge-separated states can either go on to further react or recombine to reform the ground state. Recombination of the charge-separated states is essential for reversible sensing methods. If the excited state does not undergo electron transfer the excited state will decay while emitting a lower-energy (longer wavelength) photon than the initially adsorbed photon. Most fluorophores are unable to emit a photon for all of the photons absorbed because of thermal decay pathways. Photoinduced electron transfer is simply an additional fluorescence decay pathway that can be reversible.

Proteins use distance, solvent-polarity, and reduction potentials to mediate directional electron flow as the basis for all life processes [12]. Most metal or organic cofactors buried within the protein structure facilitate electron transfer between redox-active cofactors. This is due to electron transmission being an order of magnitude higher probability in hydrophobic and organized media than water, being hydrophilic and disordered, *via* the solvation-term of the Marcus relationship [13]. Differences in redox-cofactor reduction potentials can be optimized for facile electron transfer as well, where the nuclear reorganizational energy identifies the potential difference for optimal electron transfer [12]. More important, the direction of electron flow across a multiple redox-cofactor protein can be provided by properly tuning the reduction potential differences to limit backward electron flow. Distance between redox cofactors can also be used to tune electron transfer rates and electron flow directionality. Figure 4.1 shows these different modes and relative magnitudes of effect on electron transfer rate. The final control system used by proteins in electron transfer is protein-protein or domain-domain interactions. Two examples of conformation-driven, distance-dependent electron transfer are between cytochrome *c*/cytochrome *c* oxidase [14, 15] and cytochrome *c*/cytochrome *bc*1 complex [16]. Both examples have cytochrome *c* rotate on the surface of each protein to change the distance between the heme cofactors based on the oxidation state of cytochrome *c*. Using this method directional electron flow is used in respiration and photosynthesis. In the end, the speed and direction of electron flow in biological systems are controlled by placing cofactors in organized hydrophobic media, tuning the difference in reduction potentials, and tuning and altering the distance between redox cofactors.

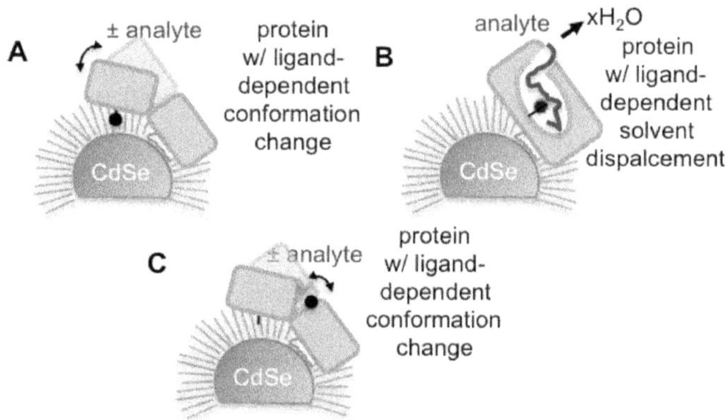

Figure 4.1 Modes that ligand dependent changes are translated into electron transfer quenching. (A) Distance; as the protein (green rectangle) conformation changes, due to ligand/analyte, the distance between the electron transfer quencher (black dot) and the quantum dot (red circle) surface is altered [27]. (B) Solvation; as a ligand/analyte binds to the protein (green rectangle) the water in the ligand binding pocket, which surrounds the electron transfer quencher (black dot), is displaced to lower the local dielectric [29]. (C) Reduction Potential; as the protein (green rectangle) conformation changes, due to ligand/analyte, the an additional ligand (orange arc) is brought into bonding distance with a electron transfer quencher (black dot) to lower the solvent dielectric and change the reduction potential.

Emission intensities of semiconductor nanoparticles, or quantum dots, are affected by both quantum dot surface charge and by changes in electron density. Electron transfer to and from quantum dots is known to cause changes in bulk [17] and single nanoparticle emission behavior [18]. While such changes are reasonable from a molecular standpoint, quantum dots behave as an intermediate case between molecules and materials. Certainly the material nature of quantum dots is evident in the size-dependent maxima of the emission wavelength. One formalism (band theory) simply treats the lower-lying valence band (VB) as a collection of highest occupied molecular orbital (HOMOs) and the higher-lying conduction band (CB) as a collection of lowest unoccupied molecular orbitals (LUMOs). Therefore, the size dependent change in maximal emission wavelength of nanoparticles is simply due to a depletion of LUMOs from the CB and HOMOs from the VB of the bulk semiconductor material. As quantum dot diameters decrease from the bulk material (zinc blende, CdSe) to diatomic CdSe, the energy of the emitted photon will increase and blue-shifting the color. Electron transfer to the CB or electron removal from (hole transfer to) the VB will both change the collective electron spin of the quantum dot and effectively quench the fluorescent emission. Excited states (CB electron; VB hole), or excitons, are formed by illuminating the ground state quantum dot to facilitate electron and/or hole transfer. Excitons are both good reductants (CB electron) and good oxidants (VB hole) and are capable of reversible electron and hole transfer. If a molecule that is reduced or oxidized by a quantum dot exciton while being held at the quantum dot surface the emission intensity should be quenched. Upon removal of the reducible or oxidizable molecule from the quantum dot surface the emission intensity should be restored due to the reversible, cyclic nature of this distance dependent electron transfer. Thus, a sensing strategy can use alteration in a redox-active molecules reduction potential or distance to the nanoparticle.

A variety of reagents can facilitate electron transfer quenching of quantum dots. Most reagents are amphipathic, in that there are polar and hydrophobic sides of the molecules. Primarily, one of the oxidation states provides a positively charged reagent, to interact with the negatively charged quantum dot surface surrounded by the hydrophobic monolayer of capping ligands. The capping ligands, which confer quantum dot water solubility, are amphipathic as well; a hydrophobic thiol on one end to coordinate the quantum dot surface and polar end to interact with water. What forms is a hydrophobic region where quenching reagents can more easily transfer electrons to the quantum dot surface. Reduction potentials, however, provide the best metric for organizing these ET quenching reagents (Fig. 4.2). (The estimated reduction potentials of CBs and VBs from popular quantum dots [19] are included for context.) The large variation in reduction potentials that provide effective ET quenching of CdSe quantum dots is surprising. The broad range in reduction potentials appears to indicate either a broad range in nuclear reorganizational energies or VB/CB reduction potentials. Neither option is reasonable. A more plausible option is that there are a variety of hole/electron acceptor orbitals in between the bandgap (termed surface states) [20] with a broad range of reduction potentials between the reduction potential of these reagents and the quantum dot VB/CB.

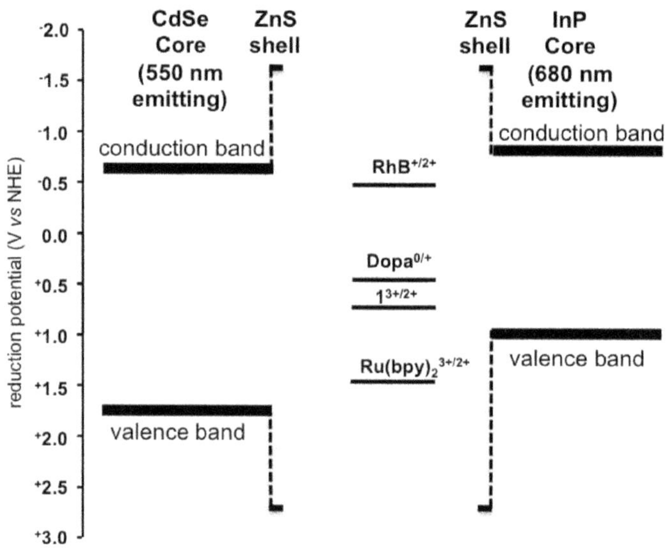

Figure 4.2 Estimated reduction potentials of quenchers at neutral pH in relation to the band structure of reported quantum dots: $\mathbf{1}^{2+}$, tetraamminephenanthroline-ruthenium(II) cation; Ru $(bpy)_3^{2+}$, trisbypyridylruthenium(II) cation; Dopa^0, dopamine; RhB$^+$, rhodamine B cation. Bandgaps are estimated from bulk materials [19] and extensions from quantum confinement. The values for ZnS shells are shown simply as the bulk material bandgap.

Quantum dot surface electronic structure is a necessary component in understanding the broad reduction potential dependence of ET quenching reagents. This simple story of quantum dot size-dependent energy separation between the top of the VB and bottom of CB (bandgap) was built on an extended or infinite lattice model. Since the quantum dot diameter is significantly smaller than the diameter an exciton typically extends in a bulk semiconductor, the orbitals generated by the termination of the lattice must be considered. While the surface states generated by lattice termination have a variety of energies, the states that should be considered in electron transfer quenching have energies within in the bandgap.

Involvement of surface states in quantum dot emission intensity is well established, but the specific effects have only been shown recently [20]. Single molecule correlation of fluorescence emission intensity and lifetime profiles has shown that a variety of emission lifetimes are obtainable from a single nanoparticle [21, 22]. This explains the multiexponential (power law) behavior of on and off emission times by single CdSe nanoparticles [23]. The manifold of emission lifetimes has been explained electron transfer to and from a surface state in the bandgap [18, 21]. Electron transfer quenching operates similarly where electrons or holes can be transferred to the surface state manifold to optimize electron transfer rates. Once an electron or hole is transferred to the surface state, the electron density can relax to the nanoparticle valence band rapidly. This opens the possibility of electron transfer with the ground state as well as with the excited state. The option of unpaired electron/hole density in a surface state or nanoparticle core presents a situation where a thermal distribution would exist (Fig. 4.3). Therefore, ET may not completely quench quantum dot emission intensity since unpaired electron/hole density in a surface state gives a variety of emission lifetimes and intensities [22]. Such a fluorescence model can be used to demonstrate this effect, which comes from a ground state (static) quenching model (Fig. 4.3B). The residual emission intensity from high quencher concentrations in a Stern–Volmer plot provides a metric for the type of surface states in thermal equilibrium with the quenched nanoparticle core.

Thickness of the quantum dot shell is, therefore, strongly controls the function of an electron transfer bio- or chemosensor. For example, thiophenol was used as an electron transfer quencher to determine ZnS shell thickness in the CdSe@ZnS nanoparticles and if cracks exist in ZnS shell [24]. While wider bandgap material shells (ZnS) function as insulators, employing these shells increase emission intensity and chemical stability. The thiophenol quenching study showed that 2–4 monolayers of ZnS provide the optimal tradeoff between improving stability/intensity while maintaining the electron transfer quenching ability. Surface states in the ZnS shell were targeted in these mechanistic discussions, where 2–4 monolayers provide enough atoms for a reasonable distribution of surface states. Unpaired electron/hole density on ZnS shell surface states increases the variability in emission lifetime and intensity and increases the residual emission intensity from ET quenching. A recent demonstration of type II CdS@ZnSe nanoparticles, that have separated the location of electron and hole wavefunctions from the exciton [25], provides an interesting situation that might couple biosensing with lasing [26]. In the end, wide bandgap material shells are necessary for electron transfer quenching of quantum dots; however, it requires careful synthetic control since shell thickness over 4 monolayers renders electron transfer quenching ineffective.

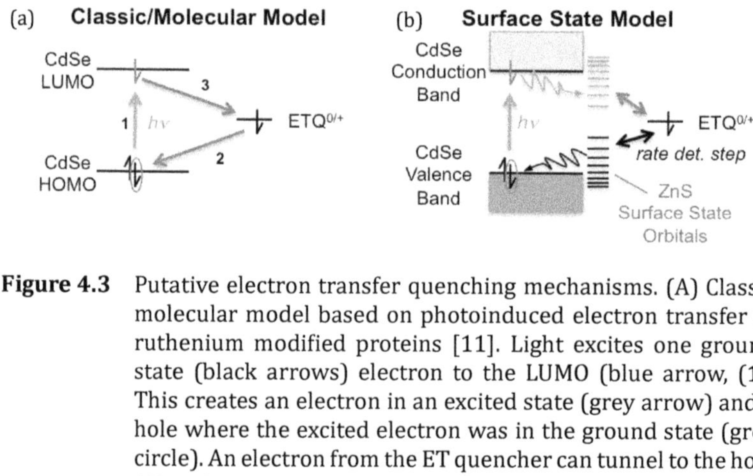

Figure 4.3 Putative electron transfer quenching mechanisms. (A) Classic molecular model based on photoinduced electron transfer in ruthenium modified proteins [11]. Light excites one ground state (black arrows) electron to the LUMO (blue arrow, (**1**). This creates an electron in an excited state (grey arrow) and a hole where the excited electron was in the ground state (grey circle). An electron from the ET quencher can tunnel to the hole (**2**) and the excited electron can tunnel to the empty orbital on the ET quencher (**3**) to quench the quantum dot fluorescence. (B) Surface state-mediated electron transfer quenching model. Reversible electron transfer between the ET quencher and a variety of surface states of the quantum dot (*rate determining step*), where the other processes outlined in the classic model are influenced by the reversible quantum dot charging (curvy arrows indicate rapid relaxation).

As a summary to this background, the combination of components for quantum dot sensor performance is critical. An amphipathic capping ligand is necessary to make a quantum dot water soluble and provide a low polarity area for the electron transfer quenching reagent to interact with the quantum dot. The ET quenching reagent must be selected for optimal interaction with the capping group/quantum dot surface. While the lower polarity environment and hydrogen bonding capability between the reagent and capping group terminus improve the electron transmission probability, it is the surface state population that can rapidly accept or donate electron density to the quantum dot. Wide bandgap shells are necessary to increase the pre-quenched emission intensities, but the thickness of these shells needs to be less than four monolayers to maintain effective electron/hole transmission from the surface state to the quantum dot core for quenching. The final component modulates the magnitude of electron transfer quenching as a function of analyte/ligand concentration.

The following sections concentrate on methods that change ET quenching using biomolecules. These methods use the basics of protein electron transfer: distance, reduction potential difference, and alteration in polarity. The modularity demonstrated for this biosensing methodology provides the potential for real-time live cell imaging of analyte concentrations.

4.2 Initial Receptor-Based Biosensors

Distance and solvation-dependent ET quenching were initially shown to provide biosensors for maltose [27, 28] and palmitate [29]. These sensors add a thiol-reactive functional group to an electron transfer quenching reagent in order to attach the reagent to the receptor. For example, [Ru(5-maleimdo-1,10-phenanthroline $(NH_3)_4][PF_6]_2$ (**1**) contains a maleimide functional group on the opposite side of phenanthroline from the Lewis basic nitrogens that coordinate the ruthenium ion [30]. A cysteine residue on the protein, which is genetically introduced, can specifically react with the maleimide functional group for site-specific placement of the ET quenching reagent. The palmitate biosensor [29] was initially designed to use changes in reduction potential to modulate electron transfer quenching; however, only minimal changes in the reduction potential of the ruthenium reagent were observed. Both of the receptors used in these biosensors have had non-quantum dot based biosensors reported previously. These systems present the best opportunity for comparison to biosensors in general.

Intestinal fatty acid binding protein (IFABP) exploits large changes in binding pocket hydrophobicity upon palmitate binding. The emission intensity of fluorescence decreases and red shifts upon palmitate binding as the fluorophore is displaced from the hydrophobic binding site, to solvent, upon palmitate [31]. A similar strategy was adopted for the first class of maltose biosensors using maltose binding protein as well [32]. It should be noted that phosphate biosensors based on the phosphate binding protein are similar to the maltose/maltose binding protein-based biosensors [33]. The Benson group showed that placing complex **1** at the bottom of the fatty acid binding site generated a protein that quenched CdSe@ZnS emission partially, but even further when palmitate was added [29]. Additional quenching was not due to an altered reduction potential of complex **1**, shown by square wave voltammetry, but fluorescence of iodoacetamidonitrobenzidiazole changes showed the polarity of the binding pocket changed with palmitate addition. The non-maleimide derivatives of complex **1** are well known to undergo solvation-dependent changes in electron transfer rate as well as reduction potential. In hindsight, the solvation-dependent change in reduction potential should not change electron transfer quenching since the manifold of surface states that the electron/hole transfers to will readjust to the new potential. In the end, solvation-mediated changes to electron transfer rate win out over changes in reduction potential as a biosensing transduction mechanism in electron transfer quenching with quantum dots.

Maltose biosensors have shown that distance-dependent changes in electron transfer quenching provide a viable biosensing transduction mechanism [27, 28]. Both CdSe and CdSe@ZnS nanoparticles were shown to function as electron acceptors in electron transfer quenching. This system anchored one domain of maltose binding protein to the nanoparticle and placed the electron transfer quenching reagent on the second domain. Once maltose binds to maltose binding protein, the domains come 10° closer to each other so that maltose is bound on the domain-domain interface. Attachment site cysteines were specifically placed away from the binding site on the surface of the domain opposite the quantum dot-attached domain. In all four attachment sites (N282C, K46C, Q72C, and K25C) maltose-dependent electron transfer quenching was observed [28]. This is significant since biosensors based on this protein superfamily with organic fluorophores require extensive testing of cysteine mutant/thiol-reactive fluorophore combinations to identify the best analyte-dependent response. In the electron transfer quenching method, all sites worked with differing amounts of quenching. Another way of stating this is that cysteines (K25C maltose binding protein) that place the electron transfer quenching reagent 20 Å away from, and on the other side of the domain from, the quantum dot attachment site still provided electron transfer quenching. This affect, that provides a modular response, arises from dynamic interactions between the quantum dot and maltose binding protein and a manifold of electron accepting surface states optimized for electron transfer quenching. So distance-dependent electron transfer quenching provided another biosensing transduction mechanism for the arsenal of maltose biosensors know to date.

4.3 Protease-Based Biosensors Show Multiplexing Ability

Three electron transfer quenching reagents were examined for use in protease biosensors. Complex **1**, succinimdylcarboxyferrocene, and (isothicyano-phenathroline)bisbipyridylruthenium(II) dication were each placed at the end of peptides that contained a endopeptidase recognition site and cysteine on the opposite end for quantum dot attachment [34]. Only the complex **1**-modified peptides yielded electron transfer quenching. Although similar compounds to the other two potential electron transfer quenchers have been shown to quench quantum dot emission [35, 36], minimal quenching was observed in this system. Admittedly the peptide linkage is the most flexible case, where complex **1** may have a higher binding affinity for the capping group/quantum dot surface than the other two reagents. More important, proteases were able to access these peptides, catalyze amide hydrolyses, and eliminate a terminal complex **1**-modified peptide to restore quantum dot emission intensity. This biosensor class was used to detect thrombin, trypsin, and chymotrypsin based on the sequence employed for the peptide linkage.

A rationale for complex **1** functioning as the only potential electron transfer quencher was suggested from electrochemical data. Square wave voltammetry of CdSe@ZnS nanoparticles showed an electrochemically active collection of surface states at a similar reduction potential to complex **1**. This electrochemical response demonstrates why a variety of electron transfer quenching reagents reported with reduction potentials in this range. The electrochemical response does not preclude molecules with different reduction potentials from functioning as electron transfer quenching reagents. As long as the molecule can associate with the capping group/quantum dot interface, the reduction potential of an electron transfer quenching reagent. For example, a DNA displacement assay with a (7,8-dimethyl-dipyrido[3,2-a:2′,3′-c]phenanthroline)bisbipyridyl-ruthenium(II) dication intercalated into the quantum dot-adsorbed duplex DNA yielded facile electron transfer quenching, which has a similar reduction potential to the (isothicyanophenanthro-line)bisbipyridylruthenium(II) dication [37]. A capping group with ferrocene terminal functional group has been shown to provide electron transfer quenching in CdSe@ZnS nanoparticles. Our analysis of this data is that the flexible peptide linkages allow for electron transfer quenching from reagents that have a significant affinity for the capping group-quantum dot surface interface.

More important, the Medintz group showed the potential for multiplexed analysis of multiple proteases in the same solution [38]. Simple monochromatic separation of emission and spectral deconvolution was able to resolve up to six quantum dots in the same solution. These results along with extensive work on capping group synthesis to provide quantum dots with high water-solubility in physiologically relevant, moderate-ionic strength solutions show great promise for employing these biosensors in live cell assays.

4.4 Recent Distance-Dependent Biosensors Demonstrate Analyte Modularity

The distance-dependent electron transfer quenching method has been extended to additional biosensor systems to demonstrate the modularity of this approach. Phosphate binding protein, mentioned above, has been used in solvation- and conformation-dependent biosensing approaches with organic fluorophores [33]. The Benson group used phosphate binding protein as a scaffold for binding site redesign to generate biosensors for Pb^{2+} ions [39]. This report integrated a variety of aspects: ligand binding site redesign, quantum dot cores from two different materials, a protein capping group, and a demonstration of the modularity provided by distance-dependent electron transfer quenching. Using a metal binding site prediction algorithm to provide a coordination sphere specific to Pb^{2+} ions, the phosphate binding site was redesigned to bind Pb^{2+} ions. The biosensors were designed with respect to monitoring lead intoxication by assaying the blood serum. To avoid the large absorbance cross-section of hemoglobin/red blood cells, the initial CdSe@ZnS-based biosensors were converted to InGaP@ZnS-based biosensors. This product is no longer sold, however InP@ZnS nanoparticles are currently commercially available (NNLabs) with similar absorbance and emission wavelengths. The mercaptohexdecanoate capping groups promoted Pb^{2+} dependent precipitation so the β-domain of metallothionein was recombinantly produced [40] and used to afford water-soluble, Pb^{2+} ion-stable quantum dots. Once these changes had been made three cysteine attachment points for complex **1** were generated on the domain opposite of the quantum dot attachment site and all three biosensor-derivatives functioned with both CdSe@ZnS and InGaP@ZnS quantum dots.

A thrombin biosensor was also generated using distance-dependent electron transfer quenching of quantum dots. This sensor simply replaced the methylene blue nucleotide modification [41] with complex **1** on the complimentary DNA strand to the thrombin binding DNA strand [42]. Upon thrombin binding to the DNA aptamer, the Plaxco and Heeger groups observed an increase in methlyene blue electrochemical current (on Au electrodes) [41] from complimentary DNA strand flexibility and an increased access of methylene blue to the Au surface. The Benson group observed the expected quench in CdSe@ZnS quantum dot emission from increased complex **1** access to the quantum dot surface [42]. Turn on sensors in the electrochemical method became turn off sensors with quantum dots, however the alternative direction should function similarly. The major hurdle to translating distance-dependent electron transfer quenching was again biosensor solubility at the salt concentrations required to maintain nucleic acid structure. Mercaptoundecylhexaethyleneglycol substitution as the quantum dot capping ligand afforded biosensor solubility and function in 1 mM Mg^{2+} ion-containing buffers. Plotting the emission response verses analyte (thrombin) concentration as a semi-log plot was essential to diagnosing this analyte-dependent precipitation/insolubility phenomenon.

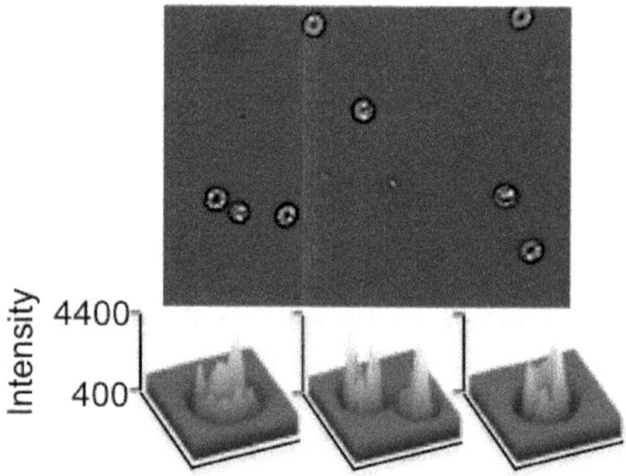

Figure 4.4 Fluorescence micrographs of rabbit red blood cells with InGaP@ZnS nanoparticles. A 7% hematocrit sample is displayed with a Xe arc lamp source, 550–650 nm excitation filter, and a 650–700 nm emission filter. Red blood cells were infected with mercaptohexadecanoate-capped InGaP@ZnS nanoparticles in the presence of streptolysin-*o*. Three dimensional graphs below the image display the log scale emission intensity of three areas that contain a red blood cell. Non-uniform emission intensity indicates carboxylate-capped quantum dot segregation within red blood cells.

Quantum dot solubility in biologically relevant buffers has become a limiting factor in applying these biosensors to fluorescent imaging of live cells. Maltose binding protein avoided these solubility issues due this protein being specifically used as a solubilization domain in protein expression. Medintz and coworkers at Naval Research Lab have been addressing solubility issues synthetically and have reported some significant advances [43]. The synthetic routes to these molecules are multistep with moderate yields that a synthetic group can easily reproduce but will require additional work for commercialization. This advance in capping group design overcomes physiological solubility from the buffer and ion perspective, however cytosolic space contains a high concentration of proteins as well. To demonstrate the insolubility issue, Fig. 4.4 shows fluorescence micrographs of metallothionein-capped InGaP@ZnS quantum dots in red blood cells. While these quantum dots were soluble in cationic, buffered solutions, the quantum dots precipitated when entering the cell via diffusion into streptolysin *o*. More than likely these negatively charged quantum dots adsorbed to cystoskelatal proteins. This result suggests that additional work is needed for live cell imaging with quantum dot-based biosensors.

4.5 Toward Biological Imaging of Analyte Concentrations

Many of the aspects needed for biosensing molecule concentrations in living cells with quantum dots have been addressed above. To recap, the shell of quantum dots, capping ligand, solubility, and biomolecule flexibility need to be carefully considered. There is one important aspect that needs to be elaborated in this venue, which is the electron transfer quenching reagent which is employed. The majority of the reported systems have used a carboxylate-terminated capping ligand to afford water solubility. It is clear that these capping ligands cannot be used effectively in live cell imaging applications. Therefore, there is a need in this field to develop additional thiol-reactive electron transfer quenching reagents. A wide variety of amphiphathic properties are needed as multiple capping groups may need to be employed for cytosolic solubility. Furthermore, electron transfer quenching reagents need to be easily synthetically accessible. One recent interesting example is the use of dopamine derivatives as electron transfer quenching reagents [44]. Table 4.1 reports molecules that have been reported to provide electron transfer quenching, primarily to CdSe@ZnS nanoparticles, which have been used or could be considered for thiol attachment to biomolecules. To date, complex **1** seems to provide the most effective electron transfer quenching potentially due to its cationic, amphipathic, and hydrogen bond donating ability. The synthetic procedures to generate complex **1** are moderate to complex, where additional options would be useful. This list of electron transfer quenching reagents hopefully shows the breadth of additional structures to consider. In the end, extensive Stern–Volmer quenching analysis will be necessary to determine a panel of robust electron transfer quenching reagents that function well with cytosolic soluble quantum dots.

Table 4.1 Electron transfer quenching reagents for quantum dots

Compound name	Structure	Conditions	Reference
1-(2-mercapto-ethyl) -3-phenyl-thiourea	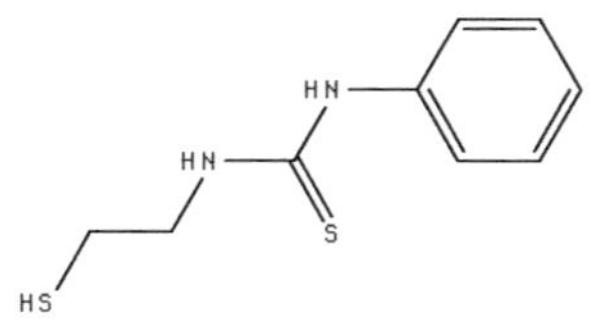	Capped on CdSe/ZnS nanoparticles in chloroform. Can detect various anions.	45
Rhodamine B	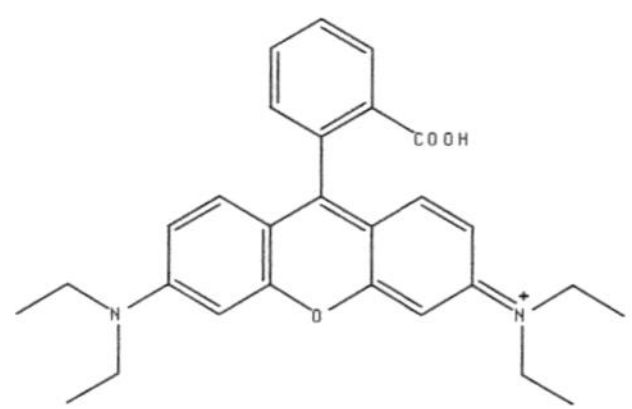	Used CdSe quantum dots capped with oleic acid in heptane	46
Ruthenium-phenanthroline maleimide	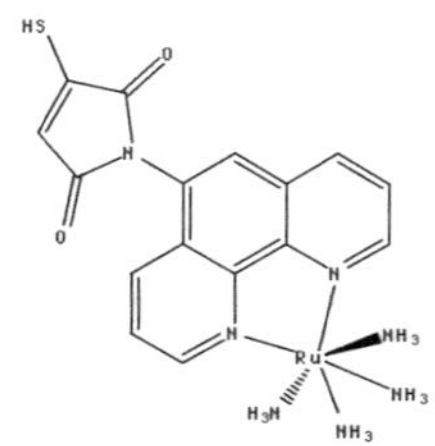	CdSe/ZnS quantum dots coated with DHLA or DHLA-PEG with peptide linker. Detects maltose when connected to maltose binding protein attached to the quantum dot. Detects palmitate when bound to intestinal fatty acid binding protein.	27

Compound name	Structure	Conditions	Reference
Ferrocene-NHS	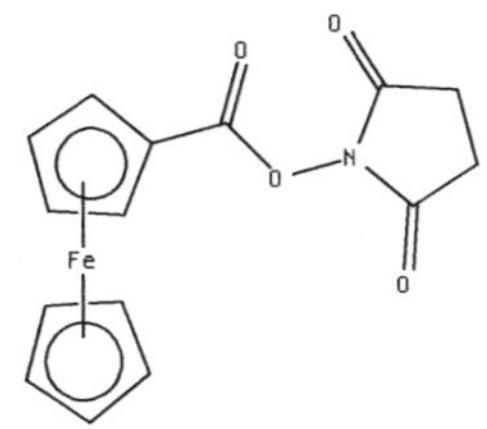	CdSe/ZnS quantum dots coated with DHLA or DHLA-PEG with peptide linker. Did not yield significant quenching.	34
Bipyridinium cation and biotin ligand	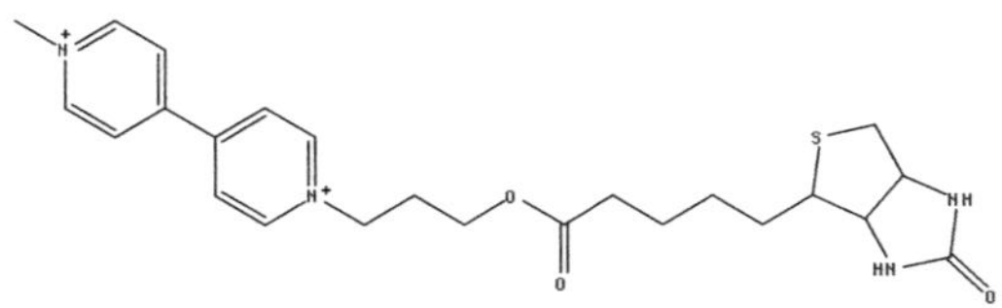	Attached to CdSe/ZnS quantum dot coated with mercaptoacetate ions. Can detect Streptavidin.	47
Phenyl boronic acid bound to nicotinamide adenine dinucleotide (NAD^+) or 1,4-dihydronicotinamide adenine dinucleotide (NADH)	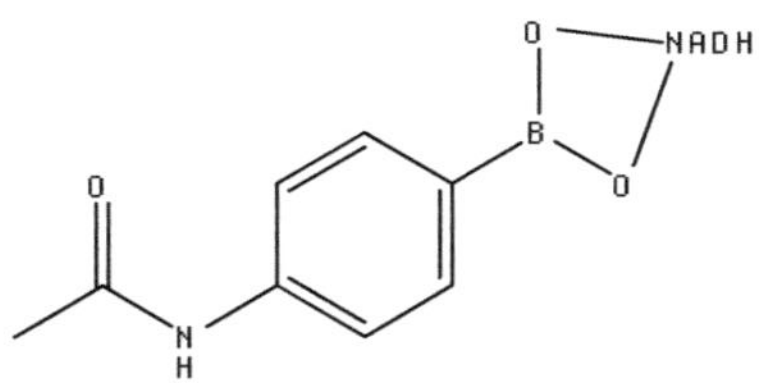	Capped onto CdSe/ZnS quantum dots. Could detect the explosive RDX.	48

(*Contd*)

Table 4.1 (*Contd*)

Compound name	Structure	Conditions	Reference
Mitoxantrone	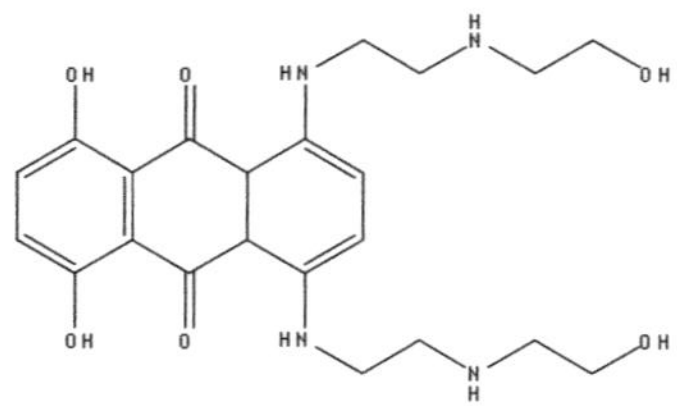	Used water-soluble CdTe quantum dots and single-stranded DNA as the analyte	49
Ruthenium bypyridine phenanthroline	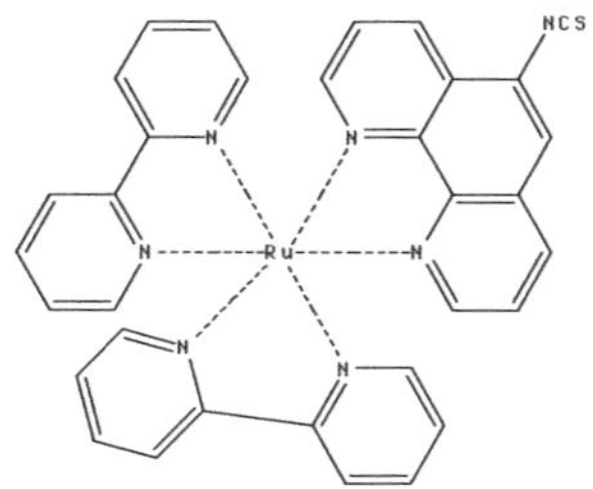	CdSe/ZnS quantum dots coated with DHLA or DHLA-PEG with peptide linker. Did not yield significant quenching.	34
L-cysteine	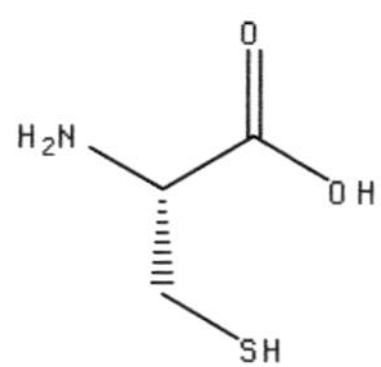	Capping group for CdS quantum dots in aqueous solution. Sensitive to zinc ions.	50

Compound name	Structure	Conditions	Reference
Thioglycerol	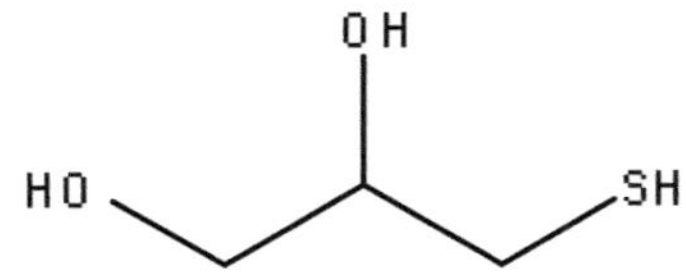	Capping group for CdS quantum dots in aqueous solution. Sensitive to copper ions.	50
Dopamine	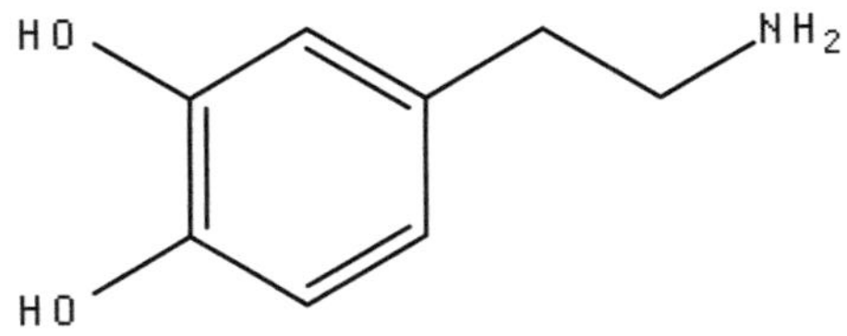	Attached to CdS:Mn/ZnS quantum dots. Could detect glutathione.	44, 51
Trinitrotoluene	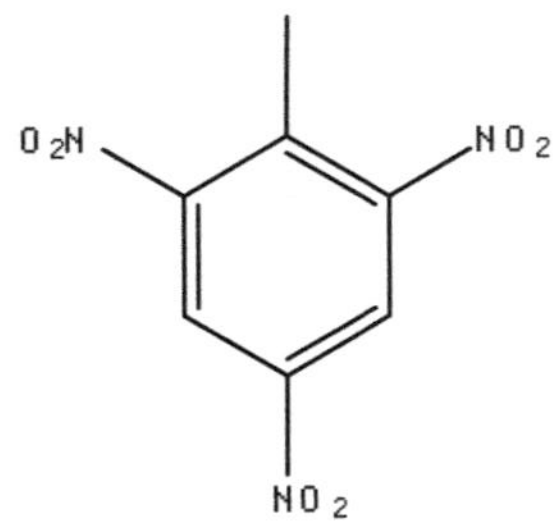	CdSe/ZnS quantum dots in toluene. Can detect trinitrotoluene with no other modifications.	52

References

1. Taube, H. Electron transfer between metal complexes — retrospective. *J. Teach. Res. Chem.*, 1994, **1**, 1–13.
2. Gray, H. B., Winkler, J. R. Electron tunneling through proteins. *Q. Rev. Biophys.*, 2003, **36**, 341–372.
3. Lewis, N. S. Chemical control of charge transfer and recombination at semiconductor photoelectrode surfaces. *Inorg. Chem.*, 2005, **44**, 6900–6911.
4. Dabbousi, B. O., Rodriguez-Viejo, J., Mikulec, F. V., Heine, J. R., Mattoussi, H., Ober, R., Jensen, K. F., Bawendi, M. G. (CdSe)ZnS core–shell quantum dots: Synthesis and optical and structural characterization of a size series of highly luminescent materials. *J. Phys. Chem. B*, 1997, **101**, 9463–9475.
5. Hines, M. A., Guyot-Sionnest, P. Synthesis and characterization of strongly luminescing ZnS-capped CdSe nanocrystals. *J. Phys. Chem.*, 1996, **100**, 468–471.
6. Gill, R., Zayats, M., Willner, I. Semiconductor quantum dots for bioanalysis. *Angew. Chem., Int. Ed.*, 2008, **47**, 7602–7625.
7. Ali, E. M., Zheng, Y., Yu, H., Ying, J. Y. Ultrasensitive Pb^{2+} detection by glutathione-capped quantum dots. *Anal. Chem.*, 2007, **79**, 9452–9458.
8. Sapsford, K. E., Berti, L., Medintz, I. L. Materials for fluorescence resonance energy transfer analysis: beyond traditional donor-acceptor combinations. *Angew. Chem., Int. Ed.*, 2006, **45**, 4562–4588.
9. Griesbeck, A. G., Hoffmann, N., Warzecha, K. Photoinduced-electron-transfer chemistry: from studies on PET processes to applications in natural product synthesis. *Acc. Chem. Res.*, 2007, **40**, 128–140.
10. Medintz, I. L., Deschamps, J. R. Maltose-binding protein: a versatile platform for prototyping biosensing. *Curr. Opin. Biotechnol.*, 2006, **17**, 17–27.
11. Winkler, J. R., Di, B., Farrow, N. A., Richards, J. H., Gray, H. B. Electron tunneling in biological molecules. *Pure Appl. Chem.*, 1999, **71**, 1753–1764.
12. Gray, H. B., Winkler, J. R. Long-range electron transfer. *Proc. Natl. Acad. Sci.*, 2005, **102**, 3534–3539.
13. Marcus, R. A., Sutin, N. Electron transfers in chemistry and biology. *Biochim. Biophys. Acta*, 1985, **811**, 265–322.
14. Flock, D., Helms, V. Protein–protein docking of electron transfer complexes: Cytochrome *c* oxidase and cytochrome *c*. *Proteins: Struct. Funct. Bioinform.*, 2002, **47**, 75–85.

15. Ostermeier, C., Harrenga, A., Ermler, U., Michel, H. Structure at 2.7 Å resolution of the *Paracoccus denitrificans* two-subunit cytochrome *c* oxidase complexed with an antibody Fv fragment. *Proc. Natl. Acad. Sci.*, 1997, **94**, 10547–10553.

16. Lange, C., Hunt, C. Crystal structure of the yeast cytochrome *bc*1 complex with its bound substrate cytochrome *c*. *Proc. Natl. Acad. Sci.*, 2002, **99**, 2800–2805.

17. Wang, C., Shim, M., Guyot-Sionnest, P. Electrochromic nanocrystal quantum dots. *Science*, 2001, **291**, 2390–2392.

18. Jin, S., Lian, T. Electron transfer dynamics from single CdSe/ZnS quantum dots to TiO_2 nanoparticles. *Nano Lett.*, 2009, **9**, 2448–2454.

19. Hagfeldt, A., Graetzel, M. Light-induced redox reactions in nanocrystalline systems. *Chem. Rev.*, 1995, **95**, 49–68.

20. Baker, D. A., Kamat, P. V. Tuning the emission of CdSe quantum dots by controlled trap enhancement. *Langmuir*, 2010, **26**, 11272–11276.

21. Zhang, K., Chang, H., Fu, A., Alivisatos, A. P., Yang, H. Continuous distribution of emission states from single CdSe/ZnS Quantum dots. *Nano Lett.*, 2006, **6**, 843–847.

22. Montiel, D., Yang, H. Observation of correlated emission intensity and polarization fluctuations in single CdSe/ZnS quantum dots. *J. Phys. Chem. A*, 2008, **112**, 9352–9355.

23. Empedocles, S. A., Neuhauser, R., Shimizu, K., Bawendi, M. G. Photoluminescence from single semiconductor nanostructures. *Adv. Mater.*, 1999, **11**, 1243–1256.

24. Vinayakan, R., Shanmugapriya, T., Nair, P. V., Ramamurthy, P., Thomas, K. G. An approach for optimizing the shell thickness of core–shell quantum dots using photoinduced charge transfer. *J. Phys. Chem. C*, 2007, **111**, 10146–10149.

25. Ivanov, S. A., Piryatinski, A., Nanda, J., Tretiak, S., Zavadil, K. R., Wallace, W. O., Werder, D., Klimov, V. I. Type-II core/shell CdS/ZnSe nanocrystals: synthesis, electronic structures, and spectroscopic properties. *J. Am. Chem. Soc.*, 2007, **129**, 11708–11719.

26. Nanda, J., Ivanov, S. A., Achermann, M., Bezel, I., Piryatinski, A., Klimov, V. I. Light amplification in the single-exciton regime using exciton –exciton repulsion in type-II nanocrystal quantum dots. *J. Phys. Chem. C*, 2007, **111**, 15382–15390.

27. Sandros, M. G., Gao, D., Benson, D. E. A modular nanoparticle-based system for reagentless small molecule biosensing. *J. Amer. Chem. Soc.*, 2005, **127**, 12198–12199.

28. Sandros, M. G., Shete, V., Benson, D. E. Selective, reversible, reagentless maltose biosensing with core–shell semiconducting nanoparticles. *Analyst*, 2006, **131**, 229–235.

29. Aryal, B. P., Benson, D. E. Electron donor solvent effects provide biosensing with quantum dots. *J. Amer. Chem. Soc.*, 2006, **128**, 15986–15987.

30. Trammell, S., Goldston, H. M., Tran, P. T., Tender, L. M., Conrad, D. W., Benson, D. E., Hellinga, H. W. Synthesis and characterization of a ruthenium(II)-based redox conjugate for reagentless biosensing. *Bioconj. Chem.*, 2001, **12**, 643–647.

31. Richieri, G. V., Ogata, R. T., Kleinfeld, A. M. A fluorescently labeled intestinal fatty acid binding protein. Interactions with fatty acids and its use in monitoring free fatty acids. *J. Biol. Chem.*, 1992, **267**, 23495–23501.

32. Gilardi, G., Zhou, L., Hibbert, L., Cass, A. E. G. Engineering the maltose binding protein for reagentless fluorescence sensing. *Anal. Chem.*, 1994, **66**, 3840–3847.

33. Brune, M., Hunter, J. L., Corrie, J. E. T., Webb, M. R. Direct, real-time measurement of rapid inorganic phosphate release using a novel fluorescent probe and its application to actomyosin subfragment 1 ATPase. *Biochemistry*, 1994, **33**, 8262–8271.

34. Medintz, I. L., Pons, T., Trammell, S. A., Grimes, A. F., English, D. S., Blanco-Canosa, J. B., Dawson, P. E., Mattoussi, H. Interactions between redox complexes and semiconductor quantum dots coupled via a peptide bridge. *J. Amer. Chem. Soc.*, 2008, **130**, 16745–16756.

35. Dorokhin, D., Tomczak, N., Velders, A. H., Reinhoudt, D. N., Vancso, G. J. Photoluminescence quenching of CdSe/ZnS quantum dots by molecular ferrocene and ferrocenyl thiol ligands. *J. Phys. Chem. C*, 2009, **113**, 18676–18680.

36. Sykora, M., Petruska, M. A., Alstrum-Acevedo, J., Bezel, I., Meyer, T., Klimov, V. I. Photoinduced charge transfer between CdSe nanocrystal quantum dots and Ru-polypyridine complexes. *J. Amer. Chem. Soc.*, 2006, **128**, 9984–9985.

37. Zhou, D., Chan, W. H., He, Z., Qiu, T. Quantum dot-ruthenium complex dyads: recognition of double-strand DNA through dual-color fluorescence detection. *Anal. Chem.*, 2009, **81**, 3537–3543.

38. Medintz, I. L., Farrell, D., Susumu, K., Trammell, S. A., Deschamps, J. R., Brunel, F. M., Dawson, P. E., Mattoussi, H. Multiplex Charge-transfer interactions between quantum dots and peptide-bridged ruthenium complexes. *Anal. Chem.*, 2009, **81**, 4831–4839.

39. Shete, V. S., Benson, D. E. Protein design provides lead(II) ion biosensors for imaging molecular fluxes around red blood cells. *Biochemistry*, 2009, **48**, 462–470.

40. Aryal, B. P., Neupane, K. P., Sandros, M. G., Benson, D. E. Metallothioneins initiate semiconducting nanoparticle cellular toxicity. *Small*, 2006, **2**, 1159–1163.

41. Xiao, Y., Piorek, B. D., Plaxco, K. W., Heeger, A. J. A reagentless signal-on architecture for electronic, aptamer-based sensors via target-induced strand displacement. *J. Amer. Chem. Soc.*, 2005, **127**, 17990–17991.

42. Swain, M. D., Octain, J., Benson, D. E. Unimolecular, soluble semiconductor nanoparticle-based biosensors for thrombin using charge/electron transfer. *Bioconj. Chem.*, 2008, **19**, 2520–2526.

43. Stewart, M. H., Susumu, K., Mei, B. C., Medintz, I. L., Delehanty, J. B., Blanco-Canosa, J. B., Dawson, P. E. Multidentate Poly(ethylene glycol) ligands provide colloidal stability to semiconductor and metallic nanocrystals in extreme conditions. *J. Amer. Chem. Soc.*, 2010, **132**, 9804–9813.

44. Medintz, I. L., Stewart, M. H., Trammell, S. A., Susumu, K., Delehanty, J. B., Mei, B. C., Mellinger, J. S., Blanco-Canosa, J. B., Dawson, P. E., Mattoussi, H. Quantum-dot/dopamine bioconjugates function as redox coupled assemblies for *in vitro* and intracellular pH sensing. *Nat. Mater.*, 2010, **9**, 676–684.

45. Callan, J., Mulrooney, R., Kamila, S., McCaughan, B. Anion sensing with luminescent quantum dots — A modular approach based on the photoinduced electron transfer (PET) mechanism. *J. Fluorescence*, 2008, **18**, 527–532.

46. Boulesbaa, A., Huang, Z., Wu, D., Lian, T. Competition between energy and electron transfer from CdSe QDs to adsorbed rhodamine B. *J. Phys. Chem. C*, 2010, **114**, 962–969.

47. Raymo, F. M., Yildiz, I. Luminescent chemosensors based on semiconductor quantum dots. *Phys. Chem. Chem. Phys.*, 2007, **9**, 2036.

48. Freeman, R., Willner, I. NAD+/NADH-sensitive quantum dots: applications to probe NAD+-dependent enzymes and to sense the RDX explosive. *Nano Lett.*, 2009, **9**, 322–326.

49. Yuan, J., Guo, W., Yang, X., Wang, E. Anticancer drug–DNA interactions measured using a photoinduced electron-transfer mechanism based on luminescent quantum dots. *Anal. Chem.*, 2009, **81**, 362–368.

50. Chen, Y., Rosenzweig, Z. Luminescent CdS Quantum dots as selective ion probes. *Anal. Chem.*, 2002, **74**, 5132–5138.

51. Banerjee, S., Kar, S., Perez, J. M., Santra, S. Quantum dot-based OFF/ON probe for detection of glutathione. *J. Phys. Chem. C*, 2009, **113**, 9659–9663.

52. Nieto, S., Santana, A., Delgado, R., Hernandez, S. P., Chamberlain, R. T., Lareau, R., Castro, M. E. In *NSTI Nanotech 2004, NSTI Nanotechnol. Conf. Trade Show*; Nano Science and Technology Institute, 2004, **3**, 399–401.

Chapter 5

Quantum Dot Probes Based on Energy Transfer Mechanisms

John F. Callan,[a,*] Bridgeen McCaughan,[a] Colin Fowley,[a] Narinder Singh,[b] Navneet Kaur,[c] and Suban Sahoo[d]

[a]*Department of Pharmacy and Pharmaceutical Sciences, School of Biomedical Sciences, The University of Ulster, BT52 1SA, Northern Ireland, UK*
[b]*Department of Chemistry, Indian Institute of Technology, Ropar, India*
[c]*Centre for Nanoscience & Nanotechnology (UIEAST), Panjab University, Chandigarh, India*
[d]*Department of Applied Chemistry, S.V. National Institute Technology, Surat 395 007, Gujarat, India*
*j.callan@ulster.ac.uk

5.1 Introduction

The impressive photopysical properties offered by semiconductor quantum dots (QDs) when compared with traditional organic dyes have seen them emerge as viable alternatives to their organic counterparts in biolabelling and sensing applications [1–3]. In particular, the broad absorption spectra of QDs, coupled with their narrow size-dependent emission spectra, mean they are ideal for use as energy donors in Förster (or fluorescence) resonance energy transfer (FRET) applications. FRET involves the non-radiative

Quantum Dot Sensors: Technology and Commercial Applications
Edited by John F. Callan and Françisco M. Raymo

ISBN 978-981-4316-00-2 (Hardcover), 978-981-4364-61-4 (eBook)
www.panstanford.com

transfer of excitation energy from an energy donor to the ground state of a nearby acceptor molecule through a dipole–dipole interaction [4]. The efficiency of energy transfer between the two fluorophores is governed by two main factors: (1) the distance between the donor and acceptor molecules and (2) a good spectral overlap between the donor's emission and acceptor's absorption spectra. In addition to these criteria, it is essential that the dipoles of both donor and acceptor molecules are parallel to each other for efficient energy transfer to occur. Should the dipoles be perpendicular the energy transfer efficiency is zero. The FRET efficiency (E) can be calculated using Eq. 5.1 [4]:

$$E = \frac{1}{1 + (r/R_0)^6} \tag{5.1}$$

where r is the actual donor-acceptor separation distance (in Å) and R_0 is the donor–acceptor separation, i.e., the distance when the energy transfer is 50% (the Förster radius). R_0 can be found from Eq. 5.2: [4]:

$$R_0^6 = \frac{9000 Q_0 (\ln 10) K^2 J}{128 \pi^5 n^4 N_A} \tag{5.2}$$

where Q_0 is the fluorescence quantum yield of the donor in the absence of the acceptor, K^2 is the dipole orientation factor (which varies from 0 for perpendicular alignment of the D-A dipoles to 4 for parallel alignment), n is the refractive index of the medium, N_A is Avogadro's number and J the spectral overlap integral that can be found from Eq. 5.3 [4]:

$$J = \int f_D(\lambda)\, \varepsilon_A(\lambda) \lambda^4 d\lambda \tag{5.3}$$

where f_D is the normalised donor emission spectrum and ε_A is the acceptor molar extinction coefficient.

Experimentally, E can be measured using either steady-state (Eq. 5.4) or time-resolved (Eq. 5.5) fluorescent experiments [5]:

$$E = \frac{1 - F_{DA}}{F_D} \tag{5.4}$$

$$E = \frac{1 - \tau_{DA}}{\tau_D} \tag{5.5}$$

where F_{DA} and F_D are the fluorescence intensity of the donor in the presence and the absence of the acceptor, respectively, while τ_{DA} and τ_D are the fluorescence lifetime of the donor in the presence and absence of the acceptor, respectively.

In spectroscopic terms, an increase in E will manifest itself in a reduction in the donor fluorophore emission with a concomitant increase in the emission from the acceptor fluorophore. From Eq. 5.1, it can be observed that there is a sixth power dependence of E on the donor–acceptor separation distance meaning that small changes in this separation can manifest itself in a significant modulation of the acceptor emission. Thus, FRET has earned the term "spectroscopic ruler" as it can be used to probe separation distances between 10 and 100 Å in biomolecules and provide information on their conformational arrangement in different environments or upon binding a target molecule [4].

Although the dipole–dipole interaction and sixth power dependence of the donor–acceptor separation distance are features of the FRET mechanism that have been determined by analysis of pairs of organic fluorophores, they have also been proven to be applicable with significantly larger QDs [6]. In addition, there are several significant benefits of including QDs as energy donors in FRET pairs:

1. The broad absorption spectra of QDs means they can be excited with any wavelength less than their emission wavelength. This flexibility of selecting an excitation wavelength that can be hundreds of nanometres less than the emission wavelength means QDs possess enormous Stokes shifts. This feature significantly reduces the "direct excitation contribution to acceptor photoemission" that is common with all organic donor–acceptor pairs due to their small Stokes shifts and broad absorption spectra [7, 8]. As a result, complicated deconvolution is often required to separate the donor and acceptor emissions thus reducing the effectiveness of the technique.
2. The narrow size tunable emission spectra of QDs means their emission wavelength can be precisely controlled through their synthesis as described in Chapter 1. This offers significant flexibility in matching the donor emission of QDs with the absorption maximum of various acceptors thus maximising the spectral overlap. It also means it is possible to match QDs with the large array of organic dye acceptors available.

3. The combination of broad absorption spectra and size tunable emission spectra mean QDs are ideal for multiplex experiments where QDs of different sizes are coupled to the same or different acceptor dyes. The narrowness of QD emission spectra with full width at half maximum (FWHM) of ~25–40 nm means that it is possible for multiple QDs to straddle a single broad absorption spectrum of an accepting organic dye.
4. The large size of QDs relative to organic dyes means it is possible to decorate the QD surface with multiple copies of acceptor molecules. The benefits of this are a significant enhancement in the overall energy transfer efficiency compared to a single donor–acceptor pair [7].

Due to these advantageous properties, QDs are finding increasing use in FRET-based sensing systems. Below, we will discuss examples where QDs have been combined with organic dyes or even other QDs to develop donor–acceptor pairs that sense for low molecular mass or biomolecules in environmental or biological samples. It would be impossible to cover all the examples of such sensors in one chapter, but a representative sample of probes in each category will be covered.

5.2 pH, Ion and Small Molecule FRET Sensors

The design strategy behind most FRET sensors involves the position or intensity of the donor's emission spectrum or the acceptors absorbance spectrum altering between the analyte free and bound states so that the energy transfer efficiency is altered [9]. Two examples of pH sensors based on this approach are worthy of mention. The first was based on the ability of the benzoxazine (Fig. 5.1a) to convert to the corresponding hemiaminal (Fig. 5.1b) upon addition of hydroxide ion to produce a CdSe/ZnS QD sensor sensitive to pH [10]. The benzoxazine was linked to the QD using a dihydrolipoic acid (DHLA) group. In acidic medium it has an absorption band centred at 375 nm, which is lost on addition of hydroxide ion and replaced by a band centred at 574 nm by its conversion to the hemiaminal. The absorbance maxima of the hemiaminal at 574 nm overlaps well with the emission wavelength of the QD (555 nm). Therefore, the QD acts as an energy donor for the hemiaminal and an estimation of solution pH determined by the reduction in QD emission. The photoluminescence (PL) of the QD–benzoxazine conjugate can be recovered upon addition of trifluoroacetic acid, clearly demonstrating the reversibility of the system. However, the requirement of chloroform as the solvent leads to obvious restrictions.

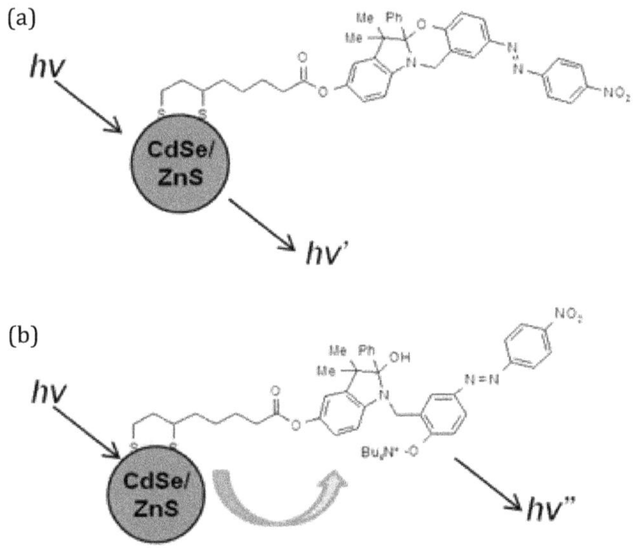

Figure 5.1 Schematic representation of (a) QD–benzoxazine conjugate which predominates at low pH and (b) QD–hemiaminal conjugate which predominates at high pH. The hemiaminal has an absorption wavelength that overlaps the QD emission and so a FRET channel is opened.

In the second example, the susceptibility of the absorption spectrum of the organic dye squarine to changes in pH was utilised by Snee *et al.* to develop a ratiometric pH sensor, operable in aqueous solution over the pH range 6–10 [11]. CdSe/ZnS QDs with their native TOPO groups present were encapsulated within an amphiphilic polymer possessing pendant alcohol groups. The squarine dye was attached by a carbodiimide mediated esterification reaction with a final QD-dye ratio of 1: 3.4 ± 0.2. The overlap between the absorption spectrum of the squarine dye and emission spectrum of the QD increased with decreasing solution pH resulting in a reduction in the QD emission with a concomitant increase in the squarine emission (Fig. 5.2). As is common with QD-dye donor acceptor pairs, ratiometric analysis is possible by comparing the ratio of the two signals or by comparing the intensity of either signal to the isoemissive point. Ratiometric analysis is preferred in biological environments because they are free from errors associated with local sensor concentration, photobleaching and environmental effects [12–13].

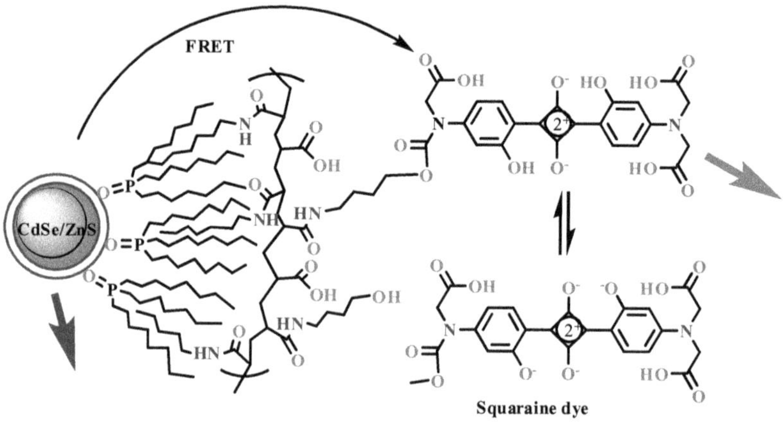

Figure 5.2 Schematic of QD–squarine dye conjugate. The native phosphine oxide ligands of the QD are encapsulated within an amphiphilic polymer upon which a pH-sensitive squarine dye is conjugated. The absorption λ_{MAX} of the squarine dye changes with changing pH and so too does its overlap with the QD emission (from Ref. 11). See also Colour Insert.

The addition of maltose binding protein (MBP) to CdSe–ZnS QDs has been used to sense for the disaccharide maltose [14] and the explosive TNT [15]. MBP was bound to a CdSe QD via an oligohistidine tag, and in the absence of maltose the fluorescence was quenched by addition of an energy-accepting non-fluorescent dye bonded to β-cyclodextrin. The cyclodextrin sugar binds to the central binding pocket of MBP, and the QD emission quenched by FRET to the dye. The addition of maltose ejects the cyclodextrin–dye complex from the binding pocket by competitive displacement and the fluorescence is restored (Fig. 5.3a). A similar strategy was employed in the detection of TNT. Anti-TNT specific antibody fragments (scFvs) were again attached to the CdSe–ZnS QDs via an oligohistidine sequence. The dye-labelled TNT analogue binds to the scFv binding/recognition site and quenches QD fluorescence by FRET as before. The addition of TNT displaces its dye-labelled analogue, the FRET channel is removed, and fluorescence is recovered (Fig. 5.3b).

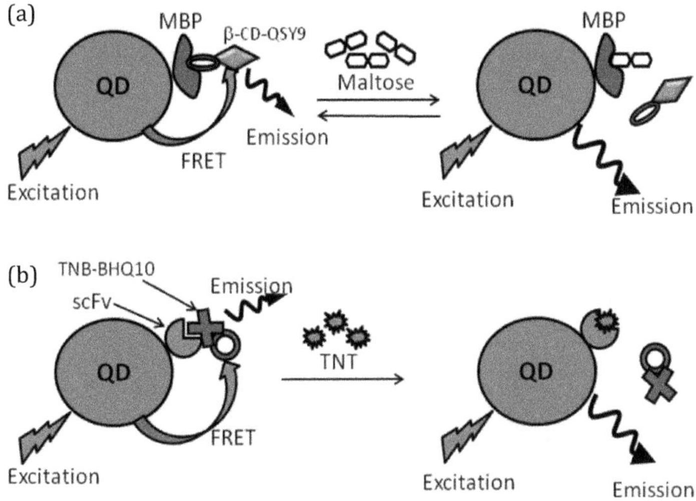

Figure 5.3 Schematic representation of a QD–conjugate sensing assembly. (a) Approximately ten MBP moieties are self-assembled on each QD; a single MBP is shown for simplicity. Formation of QD-MBP-β-CD-QSY9 results in quenching of QD emission. Added maltose displaces β-CD-QSY9 from the sensor assembly, resulting in an increase in direct QD emission. (b) Approximately 17 scFv moieties are self assembled on each QD; a single scFv is shown for simplicity. Addition of TNB-BHQ10 to QD with the TNB2-45-anti-TNT antibody fragment (scFv) quenches QD emission. As TNT is added to the assay, it completes for binding to the antibody fragment and the QD fluorescence increases following TNB-BHQ-10 release from the conjugate (from Refs. 14 and 15).

Chen *et al.* cleverly devised a CdSe/ZnS dual QD system to measure K^+ ion at the μM level in aqueous solution [16]. Two different sized quantum dots were employed: a 3.2 nm (λ_{em} = 545 nm) QD served as the energy donor and a 5.6 nm (λ_{em} = 635 nm) particle as the energy acceptor. To each was bound a 15-crown-5 receptor using a lipoic acid linker (Fig. 5.4). The crown ether showed selectivity for K^+ and formed a QD (545 nm):K^+:QD (635 nm) sandwich complex upon addition of the metal ion. Formation of the complex resulted in the different sized quantum dots coming close enough together to engage in energy transfer. A ratiometric response was therefore realised upon increasing levels of K^+, the emissions at 545 nm and 635 nm decreasing and increasing, respectively. Of course, QD (545 nm):K^+:QD (545 nm) and QD (635 nm):K^+:QD (635 nm) complexes are also possible but are statistically less probable (25% each) than the mixed size complex (50%).

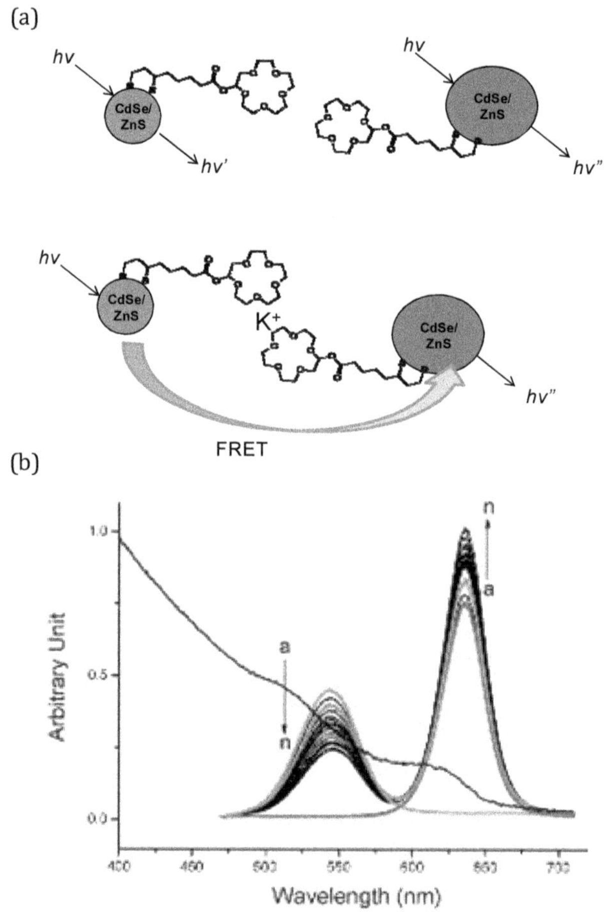

Figure 5.4 (a) Schematic representation of the QD–crown ether conjugates. When the two QDs are not complexed by K^+, they both have their own independent emission. Upon complexing K^+ they come together close enough to engage in FRET (b) spectra of the QD–crown ether conjugates upon increasing K^+ concentration: ***a*** = 0 M $KClO^{-4}$, ***n*** = 1.74×10^{-3} M.

Reudas-Rama *et al.* have developed a particulate FRET-based sensor for the extraction and dual analysis of K^+ and Cl^- ions in solution [17]. Valinomycin was utilised as a non-fluorescent selective ionophore for K^+, lucigenin as a fluorionophore for Cl^- and chromionophore I as an indicator of ion exchange and all three were incorporated with commercially available CdSe/ZnS QDs (Evident Technologies) in polyacrylate particles through a process of photoinitiated polymerisation (Fig. 5.5). The absorption spectrum of the unprotonated chromionophore I overlapped well with the emission spectrum of the QD so that FRET occurred upon excitation of the QD at 400 nm, outside chromionophore I's absorption window. The result of exciting the particles at 400 nm was dual emission at 540 nm (for the QDs) and 650 nm (for chromionophore I). Upon

(a)

(b)

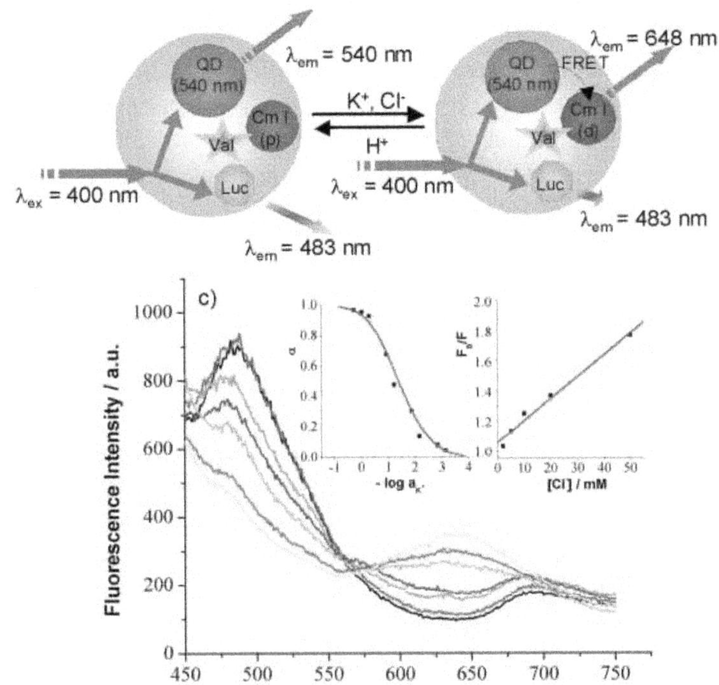

Figure 5.5 (a) Schematic of the polyacrylate beads containing the QDs, Lucigenin, Chromionophore I and Valinomycin and the changes that occur upon addition of K^+ and Cl^- ions. (b) Fluorescence spectra of beads upon addition of increasing amounts of KCl from 0 (black line) to 200 mM (yellow line) (from Ref. 17). See also Colour Insert.

the addition of K^+ ions (in the form of $KHCO_3$), the valinomycin present in the particles sequestered the K^+, which disrupted the cation proton balance resulting in a decrease in local pH. This was sufficient to protonate chromionophore I and shift its absorption spectrum significantly to the red, reducing the overlap with the QDs emission and a ratiometric change was observed in the emission (an increase in QD emission and a reduction in chromionophore I emission) upon increasing K^+ ion concentration. Furthermore, the particles were also tested for a response against chloride ions and a quenching in the emission band of lucigenin at 483 nm was observed upon increasing Cl^- ion (as NaCl) concentration. Combining the sensing qualities of the sensor for both K^+ and Cl^- through the addition of KCl, the simultaneous estimation of K^+ and Cl^- was possible using dual band detection at 650 nm and 483 nm, respectively. Although this work has some limitations, primarily given the difficulties in guaranteeing the distance between chromionophore I and the QDs in the particle, i.e., reproducing the same particles time and time again, it does highlight the significant flexibility QDs offer in selecting excitation wavelengths and positioning emission bands to enable multiplex detection.

More recently, Medintz and co-workers designed a CdSe/ZnS QD biosensor for the detection of Ca^{2+} ions in aqueous buffer (Fig. 5.6) [18]. A Ca^{2+} ion organic probe (CalciumRuby-Cl) was attached to the QD using a histidine-rich peptide sequence via metal affinity interactions. The CalciumRuby-Cl is a known photoinduced electron transfer (PET) sensor for Ca^{2+} ions where binding of the target ion to its iminodiacetate receptor results in a blocking of the PET channel to the fluorophore and a concomitant increase in emission intensity [19]. As a result, on its own, the CalciumRuby-Cl gives a single wavelength response to changes in Ca^{2+} ion concentration. However, in this study, the QD was used to indirectly excite the CalciumRuby-Cl using a wavelength (350 nm) where the dye absorbs weakly. Thus, in essence, the QD can serve as a calibrating fluorophore enabling ratiometry. In this regard, QDs have great potential in conveniently enabling ratiometric measurements for the vast array of single wavelength sensors (such as the PET family of sensors) available [20]. However, in our opinion, the real attraction of this work was the combination of both energy and electron transfer mechanisms operating within a single probe.

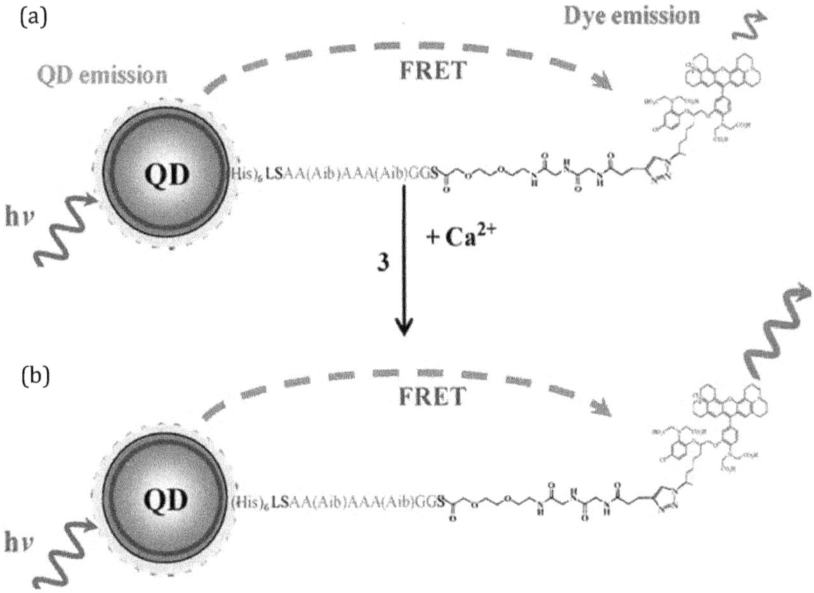

Figure 5.6 (a) Schematic representation of the QD-peptide-Calcium Ruby-Cl Ca^{2+} ion sensor. The CalciumRuby-Cl unit was indirectly excited by the QD enabling ratiometric measurement for an otherwise single wavelength probe (from Ref. 18). See also Colour Insert.

The detection of narcotics is an important area of research for a variety of reasons; so it is encouraging to see methods involving QDs being developed in this area. Zhang and Johnson have constructed both On-Off and Off-On QD probes for the detection of cocaine (Fig. 5.7) [21]. In the On-Off approach, a cocaine aptamer was sandwiched by a 3′ biotinylated oligonucleotide and a CY5-labelled oligonucleotide, which was then attached to the surface of a CdSe/ZnS QD through biotin–streptavidin binding. In the absence of cocaine, the excitation of the QD resulted in the sensitised excitation of the Cy5 dye by FRET. Upon addition of cocaine, a cocaine–aptamer complex was formed which resulted in a dissociation of the Cy5 labelled oligonucleotide from the QD-aptamer assembly thereby removing the FRET channel. In the Off-On approach, the 3′ biotinylated oligonucleotide was also labelled at its 5′ end with a Cy5 dye and the aptamer was sandwiched between this and another oligonucleotide labelled with the quencher Iowa Black RQ. The design was such that two FRET channels were possible, one between the QD and Cy5 and the other between the Cy5 and quencher. Thus in the absence of cocaine, QD excitation led to the indirect excitation of the Cy5 which was non emissive because of FRET to the quencher dye. Upon binding cocaine however, the quencher labelled oligonucleotide dissociated from the aptamer and Cy5 emission was restored.

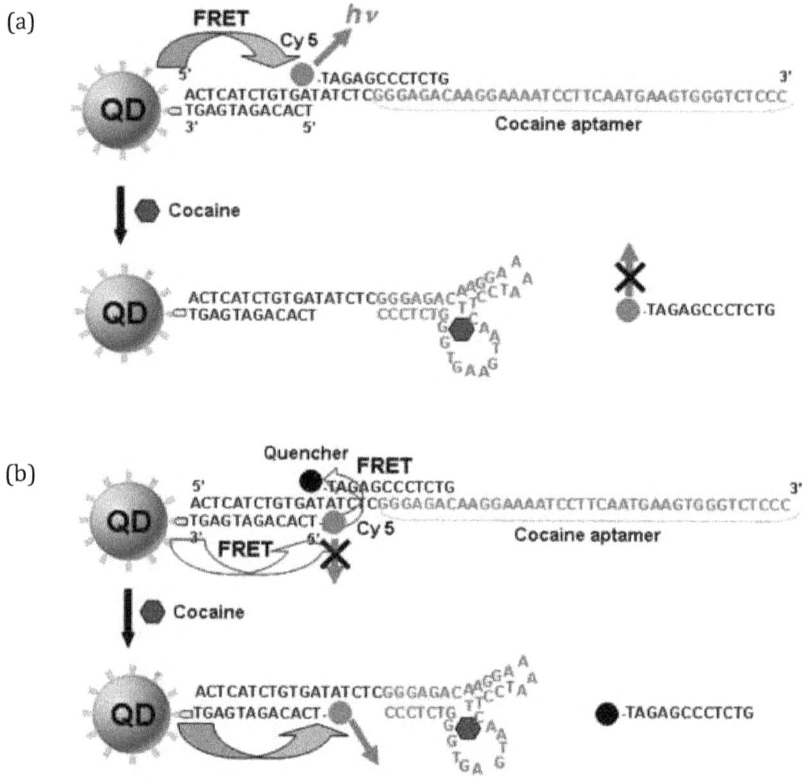

Figure 5.7 Schematic representations of (a) the On-Off and (b) the Off-On QD-aptamers for cocaine (From Ref 21). See also Colour Insert.

In the final example of this section, Yu and co-workers developed a unique approach for the detection of Cu^{2+} ions (Fig. 5.8) [22]. In all the previous examples mentioned in this section, QDs were combined with either organic dyes or other QDs. However, in this study, Yu loaded QDs into luminescent phenol formaldehyde nanoparticles (PFNs), which at 300 nm were significantly larger than the QDs. Like up-converting nanoparticles, these PFNs emit (520 nm) at a shorter wavelength than they absorb (610 nm); 610 nm emitting mercaptoacetic acid functionalised CdTe QDs were attached to the surface of the PFNs by carbodimide-based coupling techniques. Upon excitation of the PFN–QD complex at 340 nm, the QDs became excited and donated their emission energy to the PFNs resulting in an increase of the PFN emission at 520 nm. The addition of Cu^{2+} ions to a solution of the PFN–QD complex, resulted in a quenching of the QD emission most likely by electron transfer from this redox active metal ion, and a change in the colour of the solution from red to green was observed. Strangely, however, no change was observed in the intensity of the PFN emission as the QD emission decreased, even though the PFN emission was due to QD sensitisation. However, the system does provide a ratiometric method for the detection of Cu^{2+} as well as naked eye determination, although the exact mechanism for its operation needs to be explored in more detail.

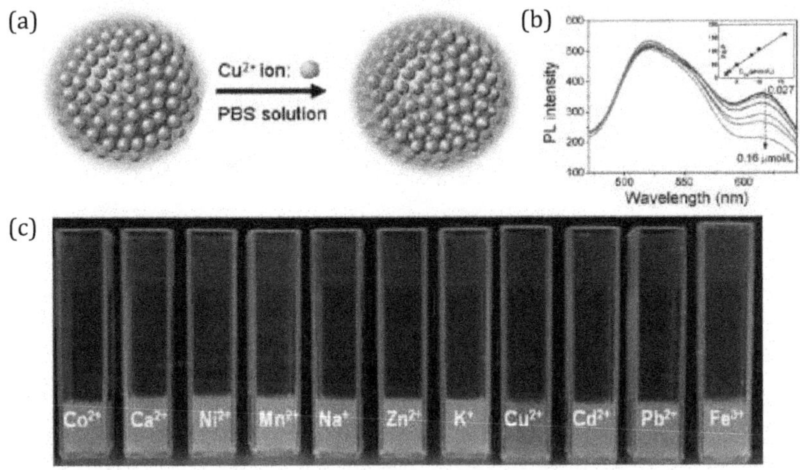

Figure 5.8 (a) Schematic representation of the PFN–QD complex with the QDs (red) adsorbed onto the surface of the PFN (green). Upon addition of Cu^{2+} ions the red QD emission is quenched and the green PFN emission is observed. (b) Fluorescence spectra of the PFN–QD complex upon addition of Cu^{2+} ions. (c) naked eye identification of Cu^{2+} ions using the PFN–QD probe (from Ref. 22).

5.3 QD-Peptide/Oligonucleotide FRET Sensors

Molecular beacons (MBs) are commonly used in biology to probe for specific nucleotide sequences. The sequence-driven complementarity of nucleotides makes them the ideal receptors in sensor design [23, 24]. Conventionally, the MB will contain an oligonucleotide sequence complementary to the sequence being targeted with a fluorophore attached at the 5′ end and a quencher dye at the 3′ end. In the absence of the target sequence FRET occurs between the fluorophore and quencher so that fluorescence remains low. However, upon hybridisation with the target sequence the separation distance between the fluorophore and the quencher increases so that FRET is no longer possible and fluorescence is switched on. Although organic dye fluorophores have traditionally been the favourite choice in the assembly of MBs, QDs are increasingly being used as alternatives due to their impressive photophysical properties. One early example by Ozkan *et al.* attached mercaptoacetic acid coated CdSe/ZnS QDs to the 5′ end of a MB whose 3′ end was attached to a organic quencher molecule 4-(4′-dimethylaminophenylazo)benzoic acid (DABCYL) (Fig. 5.9) [25]. The FRET efficiency of the QD/DABCYL MBs was investigated and compared with all organic MBs containing a 6-FAM fluorophore with DABCYL quencher. 3-D molecular modelling studies using AMBER force field calculations predicted the QD-quencher centre to

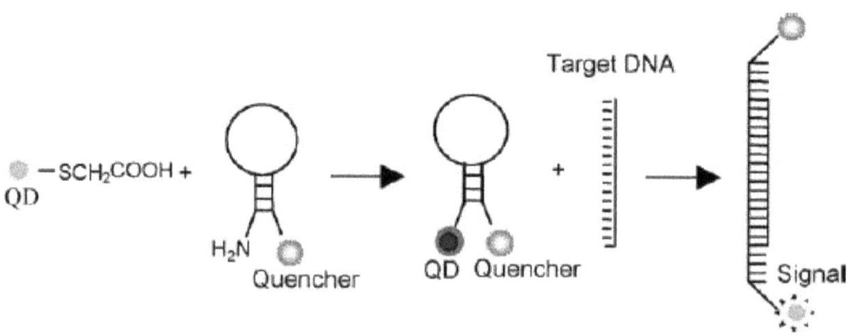

Figure 5.9 Schematic representation of the preparation of the QD MB. Upon addition of the target nucleotide the QD and quencher are distanced from each other and QD fluorescence is switched "On" (from Ref. 25).

centre distances to be slightly greater than the 6-FAM-quencher pair with a concomitant decrease in FRET efficiency. This translated to lower fluorescence enhancement for the QD MB (6 fold) compared to the all organic MB (10 fold) upon hybridization with the complementary sequence. However, a major benefit of the QD-based MBs was their high resistance to photobleaching with practically no loss in optical performance after 10 minutes irradiation compared to a 15% drop in intensity for the 6-FAM analogue.

One the limitation of conventional molecular beacons is high background fluorescence due to incomplete quenching as a result of a single donor–acceptor pair. In an attempt to overcome this, Wang and co-workers cleverly designed a quencher free FRET system based on emissive QDs and the organic fluorophore Cy5 (Fig. 5.10) [26]. Two single stranded DNA sequences complementary to the target oligonucleotide were used, one labelled with Cy5 at the 5′ end and the second with biotin at the 3′ end. These two derivatised sequences were incubated with streptavidin coated CdSe/ZnS QDs. In the absence of the target DNA strand the biotin labelled complimentary sequence can bind with the streptavidin coated QDs which cause no distortion of the QD fluorescence. However, when the target DNA sequence is present, the complementary sequences hybridise producing a hybrid with the biotin and Cy5 fluorophore at opposite ends. This hybrid can also bind to the QDs through the biotin–streptavidin interaction, and given that the emission spectrum of the QDs at 605 nm overlaps well with the Cy5 absorption spectrum, FRET occurs between the QD and Cy5 with resulting

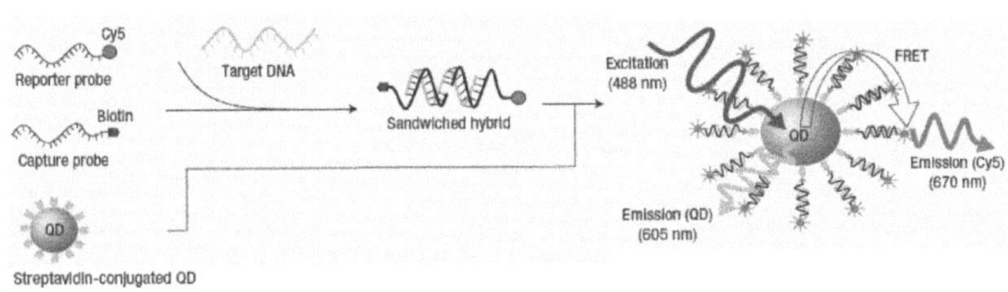

Figure 5.10 In the presence of the target DNA sequence, the capture and reporter fragments hybridise with the target sequence to form a sandwich hybrid which binds to QD–streptavidin conjugate. FRET occurs from the QD to the labelled Cy5 resulting in ratiometric emission (from Ref. 26).

decrease in QD emission and enhancement of Cy5 emission. The authors compared the effectiveness of their sensor to a commercially available Molecular Beacon and found a 100 fold increase in sensing responsivity was observed for the QD probe.

Given the similarity in size between gold nanoparticles (AuNPs) and QDs, coupled with the quenching properties of the former, it is not surprising that several groups have utilized the supramolecular association of these two distinct nanoparticles in a sensing modality. For example, Melvin and co-workers have conjugated CdSe/ZnS QDs to the 5′ end of single stranded DNA (5′-TGC AGA TAG ATA GCA G-3′) by an amide bond linkage using carbodiimide mediated coupling [27]. The complementary strand (5′-CTG CTA TCT ATC TGC-3′) was attached to an AuNP again using an amide bond linkage. When the AuNP–DNA was hybridized with the QD–DNA, an 85% quenching of the QD fluorescence intensity was observed, which could be recovered when an excess (10 equivalents) of the unlabelled strand (i.e., pure 5′-CTG CTA TCT ATC TGC-3′) was added to the mixture, as illustrated in Fig. 5.11. Although this clearly demonstrates the reversibility of the system, the time taken (3.5 hours in total) limits its use practically. However, the authors were successful in attaching the QD–DNA complex to a glass surface and observe a similar quenching effect to that observed in solution upon addition of the Au-labelled complimentary sequence.

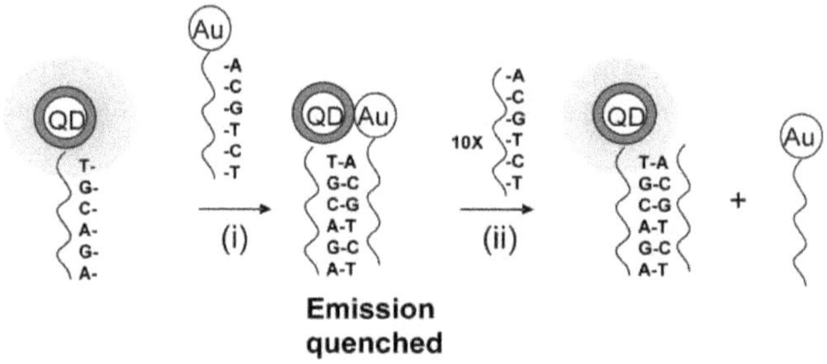

Figure 5.11 Schematic diagram showing (i) addition of 1 equivalent of the Au–DNA to the QD–DNA to yield the hybrid, (ii) addition of 10 equivalents of the unlabelled complementary oligonucleotide to the hybrid to displace the Au–DNA from the QD–DNA (from Ref. 27).

As mentioned previously, QDs are ideal for use in multiplex experiments due to their broad absorption spectra and narrow size-dependent emission spectra. Krull and co-workers exploited these advantageous properties when they used two different sized mercaptoacetic acid coated CdSe/ZnS QDs emitting in the green and red areas of the visible spectrum [28]. The green emitting QDs were attached to a nucleotide sequence complementary to the sequence diagnostic for spinal muscular dystrophy using standard EDC coupling techniques (Fig. 5.12). The larger red emitting QDs were attached to a nucleotide sequence complementary to the sequence diagnostic for *E. Coli* (Fig. 5.9b). In a slightly different approach to those outlined above, the target sequences themselves were labelled with organic dyes whose absorption spectra overlapped with those of the respective QDs (Cy3 and Alexa 647 for the green and red QDs, respectively). The addition of the labelled target sequences resulted in a concentration dependent increase in the acceptor emission due to FRET from the QD. This system enabled the detection of both target sequences simultaneously in the same solution using dual wavelength detection. However, the sensitivity of the dual nucleotide detection system relative to single nucleotide detection was significantly less, a feature attributed to non-specific absorption. Another hurdle that must be overcome when designing multiplex systems that contain more than one dye is optimizing the ratio of dyes per nanoparticle due the different fluorescence quantum yields of each dye (and possibly QD). Krull and co-workers subsequently demonstrated an improved version of their work when they immobilised the two different coloured QD-nucleic acid probes on optical fibres [29].

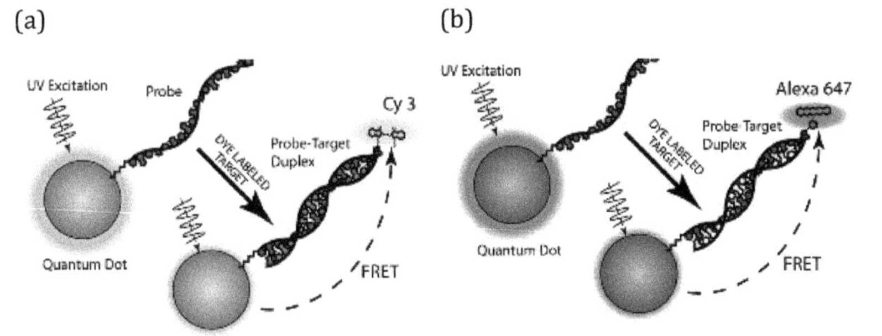

Figure 5.12 Schematic representation of (a) green and (b) red emitting QDs attached to specific nucleotide sequences. Upon addition of the labelled complimentary sequences an increase in the acceptor emission was observed due to FRET from the QD (from Ref. 28). See also Colour Insert.

5.4 QD–Antibody Conjugates

Like nucleotides, the specificity antibodies offer as potential receptors in sensing assemblies is difficult to rival with low-molecular-mass compounds. Therefore, they have been extensively used in ELISA-type detection kits [30, 31]. However, a major drawback of ELISA systems, particularly sandwich-type ELISA, is the significant washing steps required, which can prove cumbersome and time consuming. Lee and co-workers developed a QD-based "open" sandwich ELISA for the detection of oestrogen receptor b (ER-b) antigen without the need for any washing steps [32]. They attached the anti-ER-b monoclonal antibody to commercially available CdSe/ZnS QDs (emission 565 nm) using the maleimide-based cross-linker SMCC. When the ER-b antigen was added to a solution of the QD–Ab conjugate containing an Alexa Fluor dye (emission 600 nm) labelled anti-ER polyclonal antibody, the resulting QD–Ab–ER–Ab–Alexa Fluor open sandwich complex ensured FRET between the QD donor and Alexa Fluor dye (Fig. 5.13) due to the strong overlap between the QD emission and Alexa Fluor absorbance. Confocal microscope images of solutions with and without toxin clearly showed an enhancement in the Alexa Fluor emission and a reduction in QD emission for the solutions containing antigen.

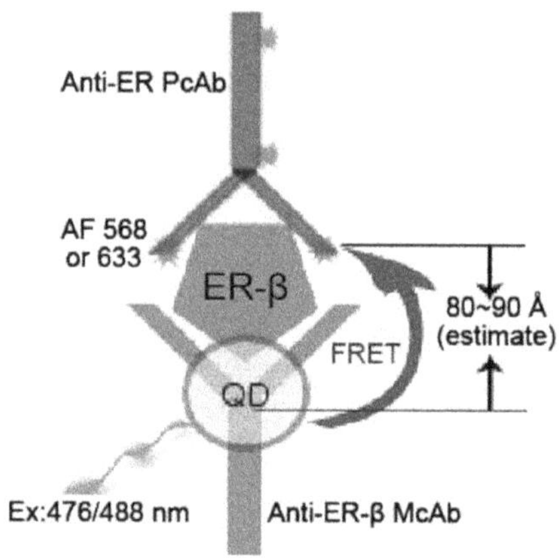

Figure 5.13 Schematic representation of the QD–Ab–ER–Ab–Alexa complex (from Ref. 32).

Although not an example of FRET, the multiplex toxin probe developed by Matoussi and colleagues will be covered here as it clearly displays the potential of QDs in multiplex analysis. They developed a sensing platform capable of detecting four toxins simultaneously using a fluoroimmunoassay approach with the aid of simple deconvolution software [33]. The wells of microtitre plates were first coated with capture antibodies raised against the cholera toxin, ricin, shiga-like toxin 1, and staphylococcal enterotoxin B. Four capture antibodies, also raised against the toxins were attached to the surface of four different sized DHLA-coated CdSe/ZnS QDs each with their own distinct emission maxima. In this sandwich ELISA approach, the presence of the specific toxins was determined by the fluorescence intensity associated with the corresponding reporter QD, once captured by their specific antibody on the microtitre plate. This method was capable of accurately measuring the four toxins in the same solution at both 1000 and 30 ng/mL concentration. This clearly illustrates the power of QDs in multiplex assays as simultaneous four analyte analysis would be non-practical with four organic dyes as each dye has its own distinct absorbance spectra with relatively poor Stokes shifts and broad emission spectra compared to QDs.

Another example of multiplex detection using an antibody-based approach, this time involving FRET in its operation, has been developed by Rogach and co-workers [34]. They incorporated two different coloured QDs (green and orange) within individual latex microbeads (MBs) which were made specific to autoantibodies by adsorption of their corresponding antigens onto their surface. Specifically, the antigen of anti-Scl-70 antibody (also called anti-topoI antibody) was adsorbed onto the MBs with embedded green QDs, while the microbeads with orange embedded QDs were coated with the anti(Sm-antigen) specific for the antinuclear antibody associated with scleroderma autoimmune diseases. To confirm the presence of either topoI or Sm antigens at the surface of the respective microbeads, monoclonal anti-topoI and anti-Sm antibodies were added, followed by labelled secondary antibodies (FITC and phycoerythrin for the green and orange QD embedded microbeads, respectively). The addition of the labelled secondary antibody opened a FRET channel from the QD to the accepting dye providing an indication of the presence of the target antibody in a flow cytometry assay (Fig. 5.14). The authors proved their system was capable of monitoring the levels of both antibodies simultaneously and illustrated the accuracy and robustness of the approach in a clinical trial by comparison of results from healthy volunteers with those having known autoimmune disease.

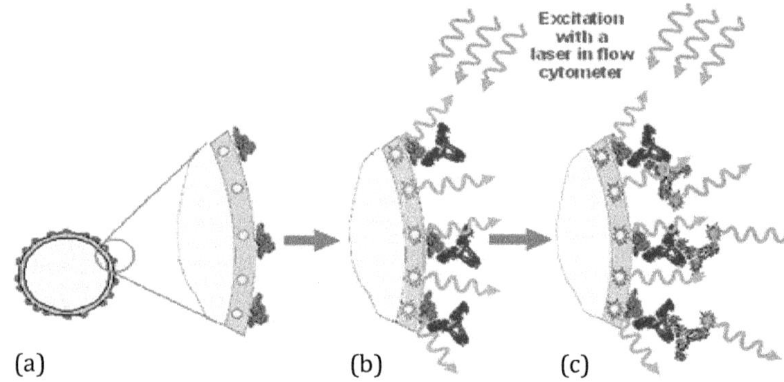

Figure 5.14 Schematic representations of the microbeads with (a) encapsulated QDs (yellow circles) and specific autoimmune antigens (green ribbons) attached to the surface (b) after addition of the unlabelled primary antibody (blue) where QD excitation results in QD emission and (c) after addition of labelled secondary antibody (brown) where QD emission leads to acceptor emission (from Ref. 34).

5.5 QD FRET Probes for Enzyme Activity

FRET probes designed for the detection of enzyme activity usually consist of a donor fluorophore attached to an acceptor fluorophore/quencher *via* bonds or specific sequences that the particular enzyme is responsible for cleaving [5]. Therefore, in the absence of enzyme, donor excitation will lead to either a quenched donor emission or acceptor emission depending on whether the acceptor is emissive or not. Not surprisingly, QDs are emerging as donor fluorophores in such systems. Rosenzweig and colleagues developed at QD-based probe for the detection of extracellular matrixmetaloproteinase (MMP) activity in normal and cancerous breast cells (Fig. 5.15) [35]. They attached a CdSe/ZnS QD to Rhodamine through a tetrapeptide sequence (RGDC) specific for MMPs. The relative emissions of the donor (QD 545 nm) and acceptor (Rhodamine 590 nm) were observed to decrease and increase, respectively, depending on the amount of the labelled RGDC peptide attached to the QD surface, suggesting an increased FRET efficiency between donor and acceptor. The addition of MMPs to the QD–peptide–acceptor conjugate led to concentration dependent increase in the donor intensity and reduction in the acceptor intensity due to cleavage of the RGDC sequence. This probe was also proved to be effective at discriminating between healthy (HTB 125) and cancerous (HTB 126) breast cell lines due to the higher MMP activity associated with the latter.

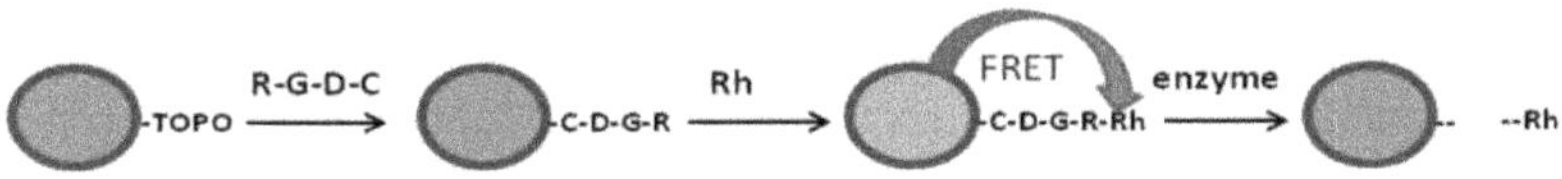

Figure 5.15 Schematic of QD–peptide conjugate labelled with Rhodamine. Enzyme addition cleaves the peptide which leads to an increase in QD emission (from Ref. 35).

Following a similar principle, Medintz *et al.* developed individual (i.e., non multiplex) QD probes to monitor caspase-1, thrombin, collagenase and chymotrypsin activity (Fig. 5.16) [8]. Peptide sequences specific for these enzymes were prepared with N-terminal His-tag sequences (for attachment to DHLA-coated CdSe/ZnS QDs through metal affinity interactions) and a C-terminal fluorophore/quencher dye (an emissive Cy3 for caspase-1 and non emissive QXL-520 for thrombin, collagenase and chymotrypsin sequences). Therefore, upon cleavage of the peptide sequence by the desired enzyme QD emission increased due to separation of the fluorophore/quencher dye from the QD. The authors preferred the use of quencher dyes over emissive dyes as acceptors due to only one signal (the QDs) observed thus removing the need of signal deconvolution. However, one must remember that having an emissive acceptor opens up the possibility of ratiometric analysis and the benefits this provides.

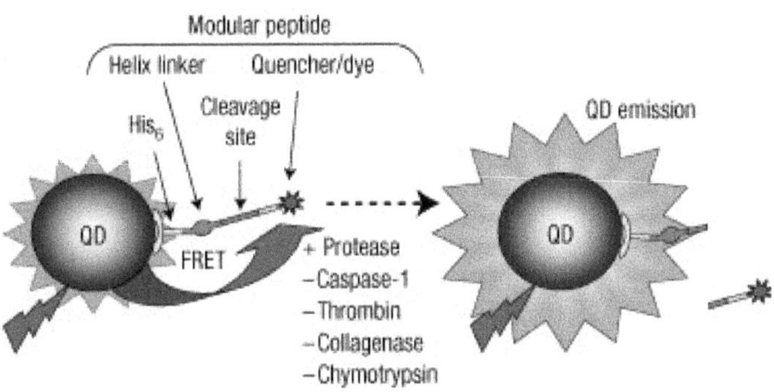

Figure 5.16 Generic schematic for the QD–peptide–label conjugate used for the detection of caspase-1, thrombin, collagenase and chymotrypsin activity (from Ref. 8). See also Colour Insert.

In another example, Rosenzweig and co-workers grew polyelectrolyte multilayers (PEMs) with incorporated CdSe/ZnS QDs in microfluidic channels enabling the real time monitoring of trypsin activity [36]. The PEMs were grown using a layer by layer deposition approach that included positively charged layers of poly(allylamine hydrochloride) (PAH) and negatively charged layers of poly(styrene sulphonate) (PSS). Negatively charged mercaptoacetic acid-coated QDs were deposited on a PAH layer and a rhodamine labelled neurotensin molecule was conjugated to the PEM film surface. Again the Rhodamine absorption overlapped well with QD emission resulting in low QD and high Rhodamine fluorescence upon excitation of the QDs in the absence of trypsin. The addition of trypsin cleaved the neurotensin, distancing the Rhodamine and QD from each other thus removing the FRET channel (Fig. 5.17). Again, the broad absorption spectra of QDs allowed the ability to choose appropriate excitation wavelengths that limit the direct excitation of organic fluorophores in FRET assays.

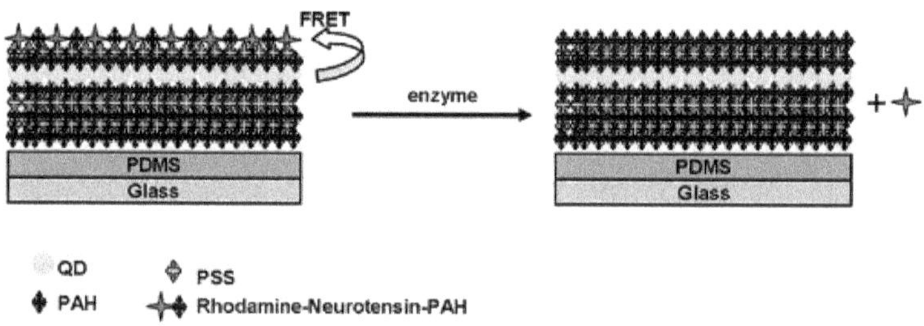

Figure 5.17 Schematic of the PEM incorporating negatively charged QDs sandwiched between two polycation layers. The outer layer of the PEM is formed from a rhodamine labelled neurotensin molecules which is cleaved upon addition of enzyme removing the FRET channel from the QD to the Rhodamine (from Ref. 36).

Another example of how AuNPs can be used as quenchers in combination with QDs is for the multiplex detection of proteases [37]. Hak-Sung Kim and co-workers attached streptavidin-conjugated QDs to an NHS-derivatised hydrogel glass slide. Monomaleimide-functionalized AuNPs were conjugated to the protease specific peptide sequence through a C-terminal cysteine residue leaving the biotin labelled N-terminal free for attachment to unreacted streptavidin sites on the immobilized QDs (Fig. 5.18a). In total, three different sized QDs were employed in an arrayed format with emission wavelengths 535 nm, 605 nm and 655 nm each connected to an individual AuNP labelled peptide sequence specific for MMP, thrombin or caspase-1. The addition of the AuNP to the immobilized QD quenched the QD fluorescence to 20% of its original level. The fluorescence was recovered in each case upon addition of the respective enzyme, again due to cleavage of the peptide substrate and distancing of the AuNP from the QD. This multiplexing was beautifully illustrated with images of probes arrayed on a glass slide with the different coloured QDs fluorescence being switched "On" upon addition of the enzymes. The process was also observed to be inhibited upon addition of enzyme inhibitors confirming the QD enhancement was indeed due to cleavage of the peptide sequence (Fig. 5.18b). The authors also noted that the initial quenching efficiency of the AuNPs was greater than commercially available organic dyes, with the added bonus that the AuNP could be used as a common energy acceptor for all three QDs, a feature not possible with organic dyes.

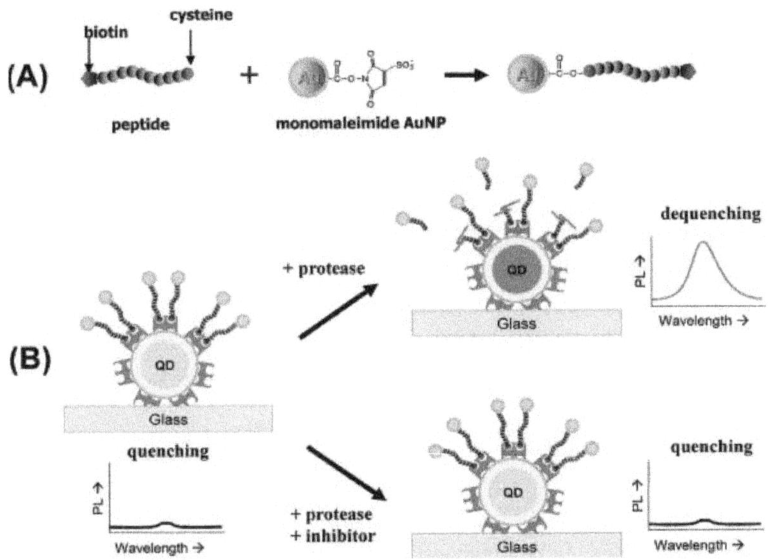

Figure 5.18 Schematic principle of protease assay by using the AuNP–QD conjugates as nanoprobes on a glass slide. (A) Construction of the peptide-conjugated AuNP; (B) Changes in the PL intensities of nanoprobes by the protease action in the absence and presence of its inhibitor (from Ref. 37). See also Colour Insert.

5.6 Conclusions

The use of QDs as energy donors or acceptors in FRET-based sensing schemes has dramatically increased in the past few years. These probes have already been used for the detection of small molecules and ions, nucleotide sequences, toxins and enzyme activity. The versatility they offer in selecting appropriate absorption and/or excitation wavelengths makes them particularly attractive for use in FRET-based multiplex experiments. Clearly, we have only yet scratched the surface in terms of potential applications for FRET operated QD probes. However, there is still much work to be done to produce nanoparticle conjugates with robust surface chemistry to enable high solubility in aqueous buffer, good stability to solution pH and a consistent batch to batch QD-ligand stoichiometry. Being able to guarantee these properties yet minimize the overall size of the nanoparticle constructs remains a challenge.

References

1. X. Michalet, F. Pinaud, T. D. Lacoste, M. Dahan, M. P. Bruchez, A. P. Alivisatos, and S. Weiss (2001), Properties of fluorescent semi-conductor nanocrystals and their application to biological labeling, *Single Mol.*, **2**(4), 261–276.
2. I. L. Medintz, H. T. Uyeda, E. R. Goldman, and H. Mattoussi (2005), Quantum dot bioconjugates for imaging, labelling and sensing, *Nat. Mater.*, **4**(6), 435–446.
3. P. Alivisatos (2004), The use of nanocrystals in biological detection, *Nat. Biotechnol.*, **22**(1), 47–52.
4. J. R. Lakowicz (2006), *Principles of Fluorescence Spectroscopy*, Springer, New York.
5. I. L. Medintz, and H. Mattoussi (2009), Quantum dot-based resonance energy transfer and its growing application in biology, *Phys. Chem. Chem. Phys.*, **11**(1), 17–45.
6. C. Curutchet, A. Franceschetti, A. Zunger, and G. D. Scholes (2008), Examining *Förste*r energy transfer for semiconductor nanocrystalline quantum dot donors and acceptors, *J. Phys. Chem. C*, **112**(35), 13336–13341.
7. A. R. Clapp, I. L. Medintz, J. M. Mauro, B. R. Fisher, M. G. Bawendi, and H. Mattoussi (2004), Fluorescence resonance energy transfer between

quantum dot donors and dye-labeled protein acceptors, *J. Am. Chem. Soc.*, **126**(1), 301–310.

8. I. L. Medintz, A. R. Clapp, F. M. Brunel, T. Tiefenbrunn, H. T. Uyeda, E. L. Chang, J. R. Deschamps, P. E. Dawson, and H. Mattoussi (2006), Proteolytic activity monitored by fluorescence resonance energy transfer through quantum-dot-peptide conjugates, *Nat. Mater.*, **5**(7), 581–589.
9. J. F. Callan, A. P. De Silva, R. C. Mulrooney, and B. Mc Caughan (2007), Luminescent sensing with quantum dots, *J. Inclusion Phenom. Macrocyclic Chem.*, **58**(3–4), 257–262.
10. M. Tomasulo, I. Yildiz, and F. M. Raymo (2006), pH-Sensitive quantum dots, *J. Phys. Chem. B*, **110**(9), 3853–3855.
11. P. T. Snee, R. C. Somers, G. Nair, J. P. Zimmer, M. G. Bawendi, and D. G. Nocera (2006), A ratiometric CdSe/ZnS nanocrystal pH sensor, *J. Am. Chem. Soc.*, **128**(41), 13320–13321.
12. G. Grynkiewicz, M. Poenie, and R. Y. Tsien (1985), A new generation of Ca^{2+} indicators with greatly improved fluorescence properties, *J. Biol. Chem.*, **260**(6), 3440–3450.
13. S. Banthia, and A. Samanta (2006), A new strategy for ratiometric fluorescence detection of transition metal ions, *J. Phys. Chem. B*, **110**(13), 6437–6440.
14. I. L. Medintz, A. R. Clapp, H. Mattoussi, E. R. Goldman, B. Fisher, and J. M. Mauro (2003), Self-assembled nanoscale biosensors based on quantum dot FRET donors, *Nat. Mater.*, **2**(9), 630–638.
15. E. R. Goldman, I. L. Medintz, J. L. Whitley, A. Hayhurst, A. R. Clapp, H. T. Uyeda, J. R. Deschamps, M. E. Lassman, and H. Mattoussi (2005), A hybrid quantum dot-antibody fragment fluorescence resonance energy transfer-based TNT sensor, *J. Am. Chem. Soc.*, **127**(18), 6744–6751.
16. C. Y. Chen, C. T. Cheng, C. W. Lai, P. W. Wu, K. C. Wu, P. T. Chou, Y. H. Chou, and H. T. Chiu (2006), Potassium ion recognition by 15-crown-5 functionalized CdSe/ZnS quantum dots in H_2O, *Chem. Commun.*, (3), 263–265.
17. M. J. Ruedas-Rama, X. Wang, and E. A. H. Hall (2007), A multi-ion particle sensor, *Chem. Commun.*, (15), 1544–1546.
18. D. E. Prasuhn, A. Feltz, J. B. Blanco-Canosa, K. Susumu, M. H. Stewart, B. C. Mei, A. V. Yakovlev, C. Loukou, J. Mallet, M. Oheim, P. E. Dawson, and I. L. Medintz (2010), Quantum dot peptide biosensors for monitoring caspase 3 proteolysis and calcium ions, *ACS Nano*, **4**(9), 5487–5497.

19. S. Gaillard, A. Yakovlev, C. Luccardini, M. Oheim, A. Feltz, and J. Mallet (2007), Synthesis and characterization of a new red-emitting Ca^{2+} indicator, calcium ruby, *Org. Lett.*, **9**(14), 2629–2632.

20. J. F. Callan, A. P. de Silva, and D. C. Magri (2005), Luminescent sensors and switches in the early 21st century, *Tetrahedron,* **61**(36), 8551–8588.

21. C. Zhang, and L. W. Johnson (2009), Single quantum-dot-based aptameric nanosensor for cocaine, *Anal. Chem.*, **81**(8), 3051–3055.

22. P. Yang, Y. Zhao, Y. Lu, Q. Xu, X. Xu, L. Dong, and S. Yu (2011), Phenol formaldehyde resin nanoparticles loaded with CdTe quantum dots: A fluorescence resonance energy transfer probe for optical visual detection of copper(II) ions, *ACS Nano,* **5**(3), 2147–2154.

23. V. V. Didenko (2001), DNA probes using fluorescence resonance energy transfer (FRET): Designs and applications, *BioTechniques,* **31**(5), 1106–1121.

24. S. Tyagi, and F. R. Kramer (1996), Molecular beacons: Probes that fluoresce upon hybridization, *Nat. Biotechnol.*, **14**(3), 303–308.

25. J. H. Kim, D. Morikis, and M. Ozkan (2004), Adaptation of inorganic quantum dots for stable molecular beacons, *Sens. Actuator B-Chem.*, **102**(2), 315–319.

26. C. Y. Zhang, H. C. Yeh, M. T. Kuroki, and T. H. Wang (2005), Single-quantum-dot-based DNA nanosensor, *Nat. Mater.*, **4**(11), 826–831.

27. L. Dyadyusha, H. Yin, S. Jaiswal, T. Brown, J. J. Baumberg, F. P. Booy, and T. Melvin (2005), Quenching of CdSe quantum dot emission, a new approach for biosensing, *Chem. Commun.*, (25), 3201–3203.

28. W. R. Algar, and U. J. Krull (2007), Towards multi-colour strategies for the detection of oligonucleotide hybridization using quantum dots as energy donors in fluorescence resonance energy transfer (FRET), *Anal. Chim. Acta,* **581**(2), 193–201.

29. W. R. Algar, and U. J. Krull (2009), Toward a multiplexed solid-phase nucleic acid hybridization assay using quantum dots as donors in fluorescence resonance energy transfer, *Anal. Chem.*, **81**(10), 4113–4120.

30. E. Engvall, and P. Perlmann (1971), Enzyme-linked immunosorbent assay (ELISA) quantitative assay of immunoglobulin-G, *Immunochemistry,* **8**(9), 871–874.

31. B. K. Van Weemen, and A. H. W. Schuurs (1971), Immunoassay using antigen-enzyme conjugates, *FEBS Lett.*, **15**(3), 232–236.

32. Q. Wei, M. Lee, X. Yu, E. K. Lee, G. H. Seong, J. Choo, and Y. W. Cho (2006), Development of an open sandwich fluoroimmunoassay based on fluorescence resonance energy transfer, *Anal. Biochem.*, **358**(1), 31–37.

33. E. R. Goldman, A. R. Clapp, G. P. Anderson, H. T. Uyeda, J. M. Mauro, I. L. Medintz, and H. Mattoussi (2004), Multiplexed toxin analysis using four colors of quantum dot fluororeagents, *Anal. Chem.*, **76**(3), 684–688.

34. A. Sukhanova, A. S. Susha, A. Bek, S. Mayilo, A. L. Rogach, J. Feldmann, V. Oleinikov, B. Reveil, B. Donvito, J. H. M. Cohen, and I. Nabiev (2007), Nanocrystal-encoded fluorescent microbeads for proteomics: Antibody profiling and diagnostics of autoimmune diseases, *Nano Lett.*, **7**(8), 2322–2327.

35. L. Shi, V. De Paoli, N. Rosenzweig, and Z. Rosenzweig (2006), Synthesis and application of quantum dots FRET-based protease sensors, *J. Am. Chem. Soc.*, **128**(32), 10378–10379.

36. G. Crivat, S. M. Da Silva, D. R. Reyes, L. E. Locascio, M. Gaitan, N. Rosenzweig, and Z. Rosenzweig (2010), Quantum dot FRET-based probes in thin films grown in microfluidic channels, *J. Am. Chem. Soc.*, **132**(5), 1460–1461.

37. Y. Kim, Y. Oh, E. Oh, S. Ko, M. Han, and H. Kim (2008), Energy transfer-based multiplexed assay of proteases by using gold nanoparticle and quantum dot conjugates on a surface, *Anal. Chem.*, **80**(12), 4634–4641.

Chapter 6

Quantum Dot Reactive Oxygen Species Generation and Toxicity in Bacteria: Mechanisms and Experimental Pitfalls

Jay Nadeau
McGill University, 3775 Rue University, Montreal, QC H3A 2B4 Canada
jay.nadeau@mcgill.ca

6.1 Overview

The toxicology of various types of nanomaterials to cells *in vitro* has been widely studied and reviewed. Key mechanisms identified include physical damage to cell structures and oxidative toxicity — both direct (e.g. lipid peroxidation, oxidation of DNA) and indirect (generation of reactive oxygen species [ROS]).[1–3] In all studies, the importance of synthesis byproducts and solvents must be carefully considered as possible confounding factors.[4,5] Other key variables that must be considered are

- Particle composition;
- Particle size;
- Surface chemistry;
- Sample aging/breakdown;
- Degree of light exposure: wavelength, power, duration.

Quantum Dot Sensors: Technology and Commercial Applications
Edited by John F. Callan and Françisco M. Raymo

ISBN 978-981-4316-00-2 (Hardcover), 978-981-4364-61-4 (eBook)
www.panstanford.com

Apart from sample considerations, it is also important to choose an appropriate test system and define the endpoints to be tested: cell death, membrane permeability, apoptosis, interference with respiration, and so on.

Given the complexity of each of these elements, it is not surprising that there is a good deal of contradiction and controversy in the nanotoxicology literature. In this chapter we will focus on the most common types of fluorescent semiconductor nanoparticles (quantum dots or QDs) in their interactions with bacteria. Many fewer toxicity studies have been done with QDs in prokaryotes as compared with mammalian cells. As of this writing, there were just over 20 published peer-reviewed articles on the subject. We will therefore be able to summarize nearly all that is known about how QDs interact with bacterial cells, and illustrate the key variables that can cause disagreements among studies, even when performed using the same bacterial strain.

To minimize the issues of QD variability, we will emphasize the most common materials and surface chemistries. We will begin with an overview of QD materials and the effects of composition on toxicity. We will then discuss ROS, the published results and the gaps in the literature, with a recommendation of which assays provide the most consistent and conclusive results. A section on photosensitization explains the ways in which singlet oxygen production may be maximized if desired. We then devote a section to the results in bacteria using three common types of QDs: core CdSe, core–shell CdSe/ZnS (or CdS), and CdTe.

6.2 Quantum Dots Cores, Shells, and Caps

The most common QD materials for biological use are made from cadmium and selenium, in the form of CdSe wurzite nanocrystals ("bare" or "core" QDs). These nanocrystals are nearly always overlaid with another semiconductor material, such as ZnS or CdS, to increase the intensity of the light emission; these are referred to as "core–shell" QDs and the outer shell may be no more than a single atomic monolayer. For CdSe, the most commonly used QD material, particle sizes of 2–6 nm show photoluminescence emission that spans the visible range; some red-shifting is seen with ZnS capping relative to the original core emission (Fig. 6.1). CdTe is another commonly used material, which can be made to emit in nearly the same colors, although its smaller bandgap means that correspondingly sized particles will be redder.

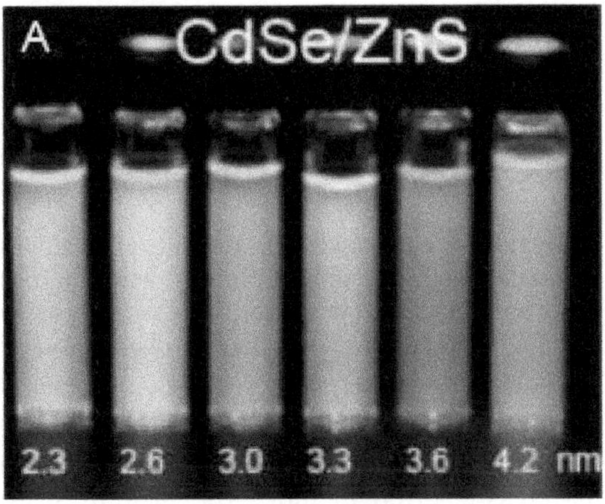

Figure 6.1 Approximately 1 μM CdSe/ZnS QD solutions in water, illuminated with a 365 nm UV wand. The numbers below the vials indicate the particle core size. See also Colour Insert.

In all quantum dot toxicity studies, it is crucial to distinguish between bare core and core/shell QDs. Breakdown of core QDs that contain cadmium leads to the release of Cd^{2+} ions and resulting massive cytotoxicity,[6] whereas breakdown of core/shell particles capped with ZnS leads only to relatively nontoxic zinc and sulfur.[7] It is important to note that nearly all biological studies using CdTe have been on bare core materials, for reasons we shall explain below. For this reason it is almost never possible to directly compare results with CdSe vs. CdTe.

QDs for biological applications are very frequently rendered water-soluble by a single-chain, functionalized thiol of varying length (mercaptoacetic acid, mercaptopriopionic acid, mercaptoundecanoic acid) (Fig. 6.2A) or by a dithiol such as dihydrolipoic acid (Fig. 6.2B). This surrounding solubilizing layer is referred to as a "cap." All of the data we will present in this chapter relate to QDs solubilized with single-chain thiols as shown in Fig. 6.2A.

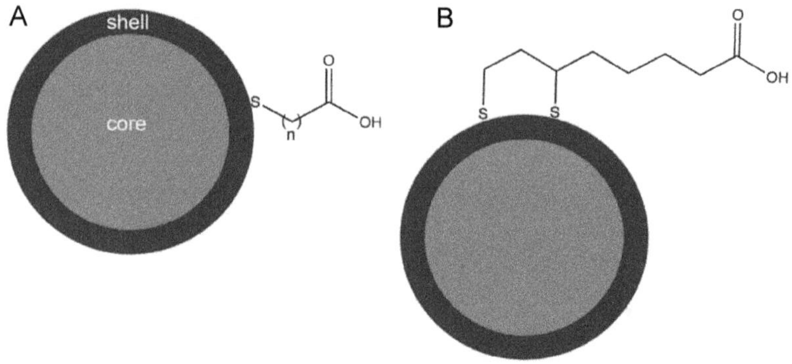

Figure 6.2 Schematic of core/shell QDs solubilized by thiols. (A) Single-chain alkanethiols: mercaptoacetic acid, mercaptopropionic acid, etc. Other simple thiols (more rarely used) are cysteine and mercaptosuccinic acid. (B) Dihydrolipoic acid.

Commercially available quantum dots are further coated with proprietary polymer and protein layers to render them stable in water; this coating often includes the protein streptavidin. In addition, there are hundreds of different solubilization methods reported in the literature. Some synthesis methods result in QDs that are already soluble, and many others replace the hydrophobic cap or further overcoat it with dendrimers, lipids, or other molecules. These methods have been well reviewed.[8] We will not address the effects of non-thiol surface chemistry on toxicity in this chapter, except to refer the reader to several excellent publications on the topic[9–12] and to reiterate that any new synthesis or solubilization method requires independent toxicity testing.

CdSe QDs solubilized with single-chain thiols are unstable in aqueous solutions in the presence of oxygen, especially when exposed to light. This may be ambient light over the course of days, or high-power illumination over the course of seconds to minutes. Photo-oxidation leads to a brightening and widening of the QD emission spectrum with a corresponding increase in fluorescence lifetime (Fig. 6.3A,B), accompanied by a loss of solubility. It has been reported that the mechanism of these changes results first from oxidation of the solubilizing ligands with the QD as a catalyst, resulting in the formation of disulfide bonds with detachment of the thiol from the QD. Eventually the nanocrystal shell and/or core

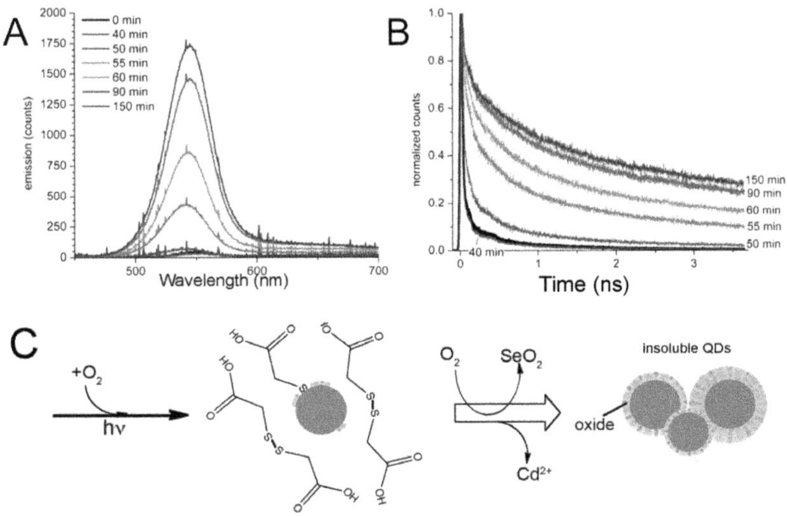

Figure 6.3 Exposure to light in aqueous solutions under air causes large increases in fluorescence and spectral broadening. (A) Emission of CdSe-MAA QDs after exposure to a UV lamp for several time intervals. Shown are the starting points after solubilization in the dark (black) and after light exposure for 40 min (violet), 50 min (blue), 55 min (green), 60 min (yellow), 90 min (red), and 150 min (maroon). (B) Corresponding time-resolved emission profiles at several time intervals, showing the starting points after solubilization in the dark (black) and after light exposure for 40 min (violet, no change), 50 min (blue), 55 min (green), 60 min (yellow), 90 min (red), and 150 min (maroon). (C) Schematic of probable changes due to light and oxygen. QDs first undergo "cap decay," where the formation of dithiols causes the capping molecules to detach from the surface. The thick lines indicate the formation of surface oxides, which increases fluorescence. As photooxidation progresses, the caps may detach completely, leaving insoluble but very bright oxide-capped particles. Different degrees of core decay lead to a broadening in the particle size distribution and thus to a broad emission spectrum as seen in Panel A.

is also photo-oxidized, leading to desorption of ions. As the size of the particle decreases, it may escape from its solubilizing cap, and precipitate from the solution. This process is called "cap decay"

(Fig. 6.3C). The component ions from the nanocrystal shell and/or core are an important cause of toxicity. Inductively coupled plasma optical emission spectroscopy (ICP/OES) measurements of ionic concentrations in solution show cytotoxic levels of free Cd^{2+} ions in solutions of bare-core CdSe exposed to UV irradiation, from 10 ppm at 1 h of irradiation to >80 ppm at 8 h. Certain types of cells, such as hepatocytes, are especially sensitive to Cd^{2+} and massive cell death can occur at these concentrations.[6] Core–shell CdSe/ZnS QDs do not release detectable Cd^{2+} in such time periods, but liberate comparable amounts of Zn^{2+}[7].

In contrast to CdSe, CdTe QDs grow *less* fluorescent with light exposure, with correspondingly faster lifetimes. It is believed that the mechanism of cap decay is the same, but the solubilizing thiols have a different effect on CdTe, enhancing the particles rather than quenching as is seen with CdSe.[13] It is for this reason that core CdTe solubilized with thiols is very bright and useful for imaging, whereas bare core CdSe is often too weakly fluorescent to be of any use.

Significant cellular damage and death are observed even when no heavy metals are released, and in cell types such as bacteria that are highly resistant to metal ions. It is these mechanisms of toxicity that have proven difficult to quantify and that vary among QD types and sizes and even synthesis batches. There is agreement among all studies that this toxicity is "oxidative," but what this means precisely is open to interpretation. In order to begin to understand this, it is important to understand the band structure of semiconductor nanoparticles and possible sources of oxidative species in aqueous solutions of these materials.

6.3 Band Edge Energies and Possible Reactions in Solution

Two physical features of a QD core determine its reactivity: the energetic position of the band edges, which is a property of the bulk semiconductor material from which the dot is made; and the size of the bandgap, which is a function of QD material and size. The band edge energies in solution correspond to redox potentials. Figure 6.4A shows these energies for several bulk QD/nanoparticle materials relative to oxygen-related redox couples, and Fig. 6.4B indicates how CdSe band edges shift with dot sizes.

It can be seen from this figure that a semiconductor may act as a reducing agent (by transferring an electron to an acceptor "A") and/or an oxidizing agent (by transferring a hole to a donor "D"). this leads to a wide variety of reactions that are possible in complex solutions containing acceptors A and donors D. The **reductive process** involves a conduction band electron and an acceptor:

$$A + e^-_{CB} \rightarrow A^{\bullet -} \tag{6.1a}$$

whereas the **oxidative process** involves the valence band hole and a donor:

$$D + h^+_{VB} \rightarrow D^{\bullet +} \tag{6.1b}$$

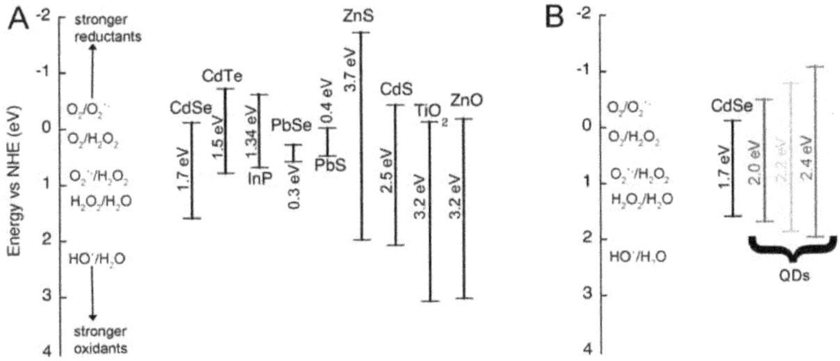

Figure 6.4 QDs as redox agents. (A) Redox potentials of oxygen species with respect to the band edges of bulk semiconductors (these may vary with pH and other factors; see ref. 69). It can be expected, for example, that the electron of photoexcited CdTe could reduce oxygen, whereas that of CdSe could not. Similarly, only TiO_2 and ZnO might be expected to generate hydroxyl radicals from water. (B) Broadening of CdSe bandgap as a function of size. Shown are the band edges for bulk CdSe and red, green, and blue QDs (emission peaks 620, 560, 520 nm). The energetic position of the electron moves approximately three times as much as the hole because of its lighter quasiparticle mass.

If A or D are water or oxygen, or if they interact with water or oxygen, then different types of ROS may result. It is important to distinguish the different types of ROS, their potential cytotoxicity, and the types of reactions that lead to their formation. Table 6.1 gives the names, structures, redox potentials, lifetimes in solution, and cytotoxic potential of several key types of ROS.

Table 6.1 Some reactive oxygen species

Species name	Species formula	Redox potential vs. NHE	Half-life in solution	Cytotoxic potential
Hydrogen peroxide	H_2O_2		Stable, but reacts with cellular enzymes	Lowest
Superoxide	$O_2^{\bullet-}$	−0.33	Minutes, but dismutated by enzymes	Moderate
Singlet oxygen	$^1O_2$		10^{-6} s	Very high
Hydroxyl radical	$OH^{\bullet}$		10^{-9} s	Extremely high

Note: For a full discussion, see refs. [65, 66].

A quick examination of the redox potentials vs. band edges gives an idea of which reactions can occur efficiently in the presence of certain QDs. For example, although bulk CdSe is not likely to reduce molecular oxygen to yield superoxide, CdSe QDs of all colors have energetic enough electrons so that the direct reaction is possible. Superoxide may then lead to singlet oxygen generation:

$$O_2 + e^-_{CB} \rightarrow O_2^{\bullet-}$$
$$O_2^- + H^+ \rightarrow HO_2,\ pK = 4.8$$
$$2HO_2\ (\text{or } 2O_2^-) \rightarrow H_2O_2\ (\text{or } O_2) + {}^1O_2 \tag{6.2}$$

Generation of highly oxidizing species, such as hydroxyl radicals, can also occur via indirect mechanisms. Direct formation is unlikely with CdSe, and even less favored with CdTe. However, the presence of metal ions can generate hydroxyl radicals via the Fenton reaction:

$$H_2O_2 + Fe^{2+} \rightarrow OH^{\cdot} + HO^- + Fe^{3+} \tag{6.3}$$

Certain species in the solution, such as catechols, may make the generation of superoxide more efficient by acting as donors in the reaction from catechol to semiquinone to quinone:

$$D + h^+_{VB} \rightarrow D^{\bullet+}\ (\text{semiquinone})$$
$$D^{\bullet+} + O_2 \rightarrow O_2^{\bullet-} + \text{quinone} \tag{6.4}$$

However, quinones in the solution may lead to *less* particle reactivity, as they can "shuttle" electrons from the conduction band back to the valence band[14]:

$$A + e^-_{CB} \rightarrow A^{\bullet-}$$
$$A^{\bullet-} + h^-_{VB} \rightarrow A \tag{6.5}$$

Because of all of these complications, it is often difficult to predict which species will be generated from QDs. In addition, the presence of a shell affects the redox reactions by creating an energy barrier through which the electrons or holes must tunnel before they can react; the size of the barrier is estimated at 4 meV for a two-atomic monolayer thickness.[15] Surface chemistry can also affect redox properties. It is likely that the difference in the valence band energy between CdSe and CdTe explains why the former are quenched by surface thiols and the latter are not; the more-energetic hole of CdSe can be trapped by a thiol.[13] Thus, for each type or even batch of QDs, it is important to measure ROS and to distinguish the types of most interest to the particular experiment. These measurements are not trivial and have generated a variety of results which we discuss in the next section.

6.4 Measuring Reactive Oxygen Species in the Presence of QDs

The apparent contradictions in the QD ROS literature can be mostly reconciled by a careful examination of the particle composition used in each study, the solvent used, and the type of ROS detected. A systematic study of free radicals generated from CdSe and CdSe/ZnS QDs was not done until 2005. Table 6.2 gives a summary of several key studies in this area. It can be seen that the data are not complete or comprehensive, as individual studies may use a solvent, surface chemistry, or type of particle that is not used by any other research group.

Table 6.2 Some studies measuring ROS generation from QDs in solution

Type of dot	Solvent	Type of ROS	Measured by	Reference
CdS	H_2O	Hydroxyl radicals, superoxide	EPR, Amplex Red, terephthalate	67
CdS, CdSe, CdSe/ZnS	H_2O	Hydroxyl radicals, superoxide (None found with CdSe/ZnS; only hydroxyl with CdSe)	EPR, SOD, terephthalate	15
CdTe, CdTe + Pc	H_2O	$^1O_2$ (low but detectable with QD only)	Emission at 1270 nm, ADMA	35
CdSe	toluene	$^1O_2$	Emission at 1270 nm	25

Type of dot	Solvent	Type of ROS	Measured by	Reference
CdTe (water synthesis)	H_2O	$^1O_2$	Emission at 1270 nm	16
CdTe	H_2O	$^1O_2$ (None found)	Emission at 1270 nm	37
CdTe	H_2O and toluene	$^1O_2$ (None found) Hydroxyl (some)	EPR	68
CdSe/ZnS	H_2O	$^1O_2$ (None found) Hydroxyl (some)	EPR, XTT	23

Abbreviations: Pc, phthalocyanine; SOD, superoxide dismutase; ADMA, anthracene-9,10-diyl-bis-methylmalonate; others as in text.

However, some conclusions can be drawn from the studies in solution. One of the most conclusive ways to measure and distinguish ROS is spin trap electron paramagnetic resonance (EPR) spectroscopy for both hydroxyl radicals and singlet oxygen. EPR, also called or electron spin resonance (ESR) spectroscopy, is analogous to nuclear magnetic resonance (NMR) except that it is electron spins rather than nuclear spins that are excited. Because a signal requires an unpaired electron, NMR is used to detect free radicals or complexes containing a transition metal ion. In spin-trap EPR, a stable radical species is provided as a "spin probe" (analagous to isotopic substitution in NMR). Changes to the spin probe alter the signal by creating or removing an unpaired electron. For example, 2,2,6,6-tetramethyl-1-piperidinyloxy (TEMPO) is a stable free radical that can be oxidized by holes, OH radicals, or any other oxidative species that have a redox potential $\geq$ +0.75 V vs. NHE. Thus the disappearance of TEMPO radical EPR spectra upon irradiation indicates photogenerated oxidative species (either hydroxyl radicals or holes) (Fig. 6.5A). For reductive species detection, the TMP method can be used. TMP is 2,2,6,6-Tetramethylpiperidine and is initially EPR-silent, but reaction with singlet oxygen/superoxide yields a stable, EPR- sensitive radical adduct. In this case it is thus the formation of the radical rather than its disappearance which is measured, and the kinetics of formation can give a clue to the mechanisms (Fig. 6.5B,C). Another spin trap commonly used is 5,5-dimethyl-1-pyrroline-*N*-oxide (DMPO), which can indicate both oxidative and reductive species by the formation of different adducts. Singlet oxygen may also be detected by direct measurement of phosphorescence emission at 1270 nm; however, this is weak and short-lived, and a positive control is necessary to ensure that the species is being detected.

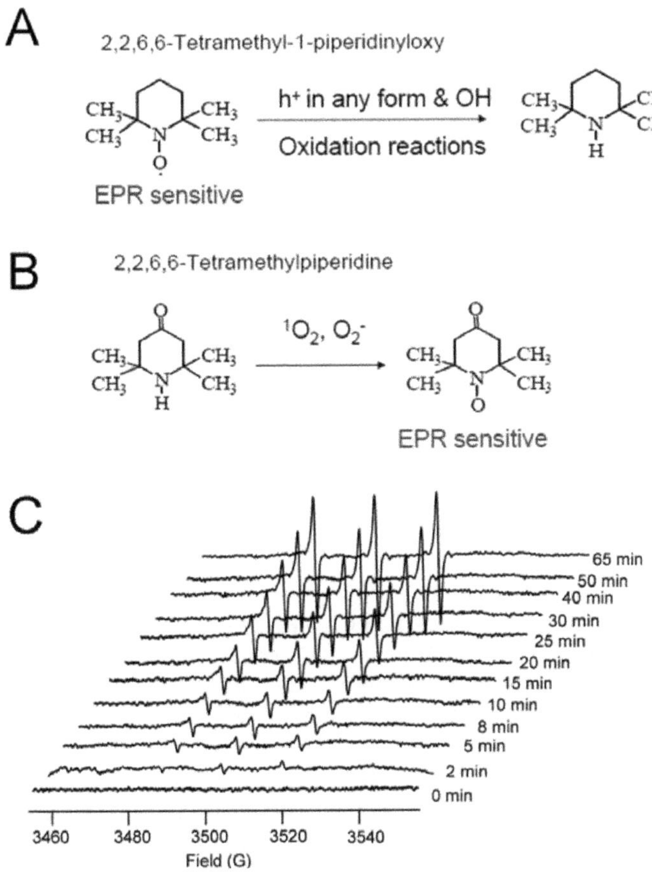

Figure 6.5 Spin-trap EPR spectroscopy for detection of reductive and oxidative species. (A) TEMPO. (B) TMP. (C) Evolution of TMP signal with time.

All of the studies cited in Table 6.2 used one or both of the above techniques, and some firm conclusions can be drawn. Highly reactive CdS nanoparticles are capable of generating large amounts of ROS. Bare core CdSe can generate small amounts of hydroxyl radicals but not substantial reducing species. CdSe/ZnS does not make superoxide but may make small amounts of hydroxyl radical; this is probably a function of the cap thickness and/or surface cap. With CdTe the situation is somewhat murkier. One study has seen some singlet oxygen from unmodified CdTe, but this was using very small particles synthesized in aqueous solution, which are probably very different from those made in organic solvent and then solubilized.[16] Other studies found no singlet oxygen with CdTe. Measurement of hydroxyl radical generation from CdTe has very rarely been done; we saw a significant signal using spin-trap EPR using TEMPO. This was seen only with thiol-capped CdTe in H_2O; no signal was apparent in toluene (Fig. 6.6A). We observed a similar signal with CdSe/ZnS, although the disappearance of the TEMPO radical was slower than with CdTe (Fig. 6.6B).

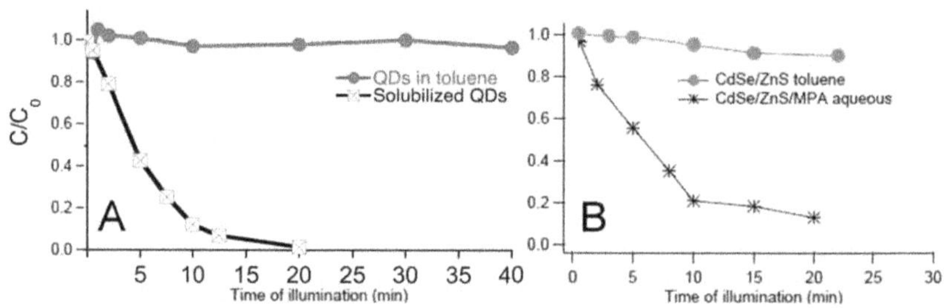

Figure 6.6 Results of TEMPO EPR using (A) CdTe and (B) CdSe/ZnS in toluene and water. QD concentrations were ~1 μM. C/C_0 indicates the concentration of the spin probe remaining relative to its initial value (typically 33 μM).

Many of these studies confirm the EPR or fluorescence observations with indirect measurements of ROS using reporter dyes. It is important to be prudent when using these indirect measures, especially with nanoparticles that may take part in a variety of reactions. A recent study found that some fluorescent reporters that generate a signal when oxidized gave false positive ROS results with fullerenes.[17] These included the commonly used probes 2′,7′-dichlorodihydrofluorescein diacetate (H_2DCFDA) and hydroethidine. Only reporters that required reduction to generate the signal were reliable, notably the XTT assay. 2,3-bis(2-methoxy-4-nitro-5-sulfophenyl)-2H-tetrazolium-5-carboxanilide (XTT) is reduced by superoxide to XXT-formazan, which can be dectected colorimetrically. The XTT assay has been used in a few studies to confirm EPR results using QDs, and does not appear to pose any consistency problems for either positive or negative results (Fig. 6.7A).

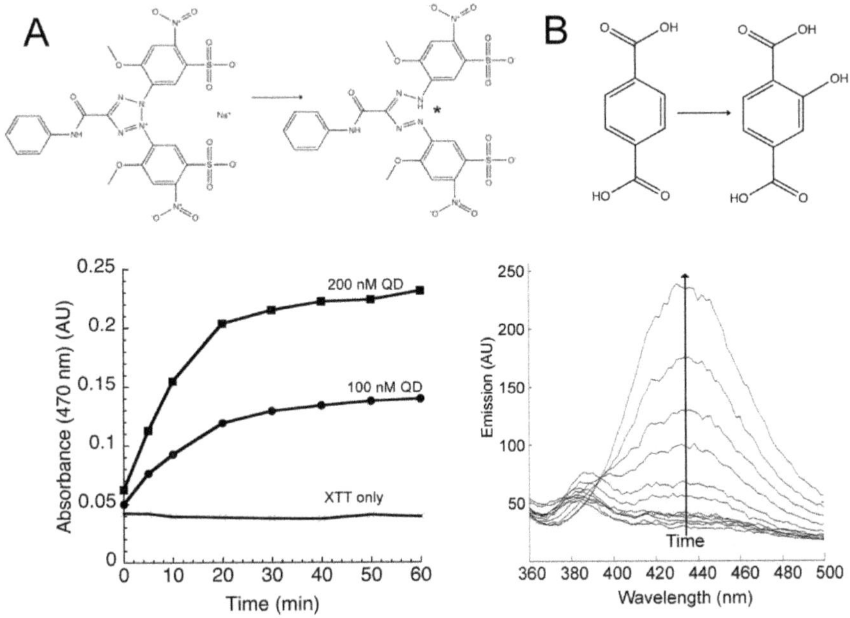

Figure 6.7 Colorimetric or fluorescent ROS reporters that work well with QDs. (A) XTT is reduced to XTT formazan, which is red (the star indicates the location of the bond change). The signal with CdSe/ZnS is dependent upon QD concentration and duration of blue light irradiation. (B) Consistent results are obtained using terephthalate, which is oxidized by hydroxyl radicals to a fluorescent product. The graph shows the fluorescence emission with time of blue light exposure. Excitation at 350 nm.

Sodium terephthalate has also been used to detect hydroxyl radicals. Terephthalate is oxidized to 2-hydroxyterephthalate (OH-TA), which shows a brilliant blue fluorescence (Fig. 6.7B). We have also obtained consistent results using singlet oxygen sensor green (SOS-G),[7] whose exact structure has not been published but which is presumed to be a fluorescein bound to a dimethylanthracene derivative. SOS-G emits with a peak at 525 nm when excited at 504 nm (UV light will bleach it and should be avoided), and does not show a signal in the presence of hydroxyl radicals or peroxide. We have found that it does not show a measurable signal with any of our solubilized QDs, as expected.

6.5 Pitfalls: Dyes to Use with Caution

The use of dihydroethidine with QDs has been reported in at least one study using cells,[18] but it is not known if the cell results were consistent with the amount and types of ROS seen in solution. H_2DCFDA has also been used,[19–22] but we have found serious issues with this particular reporter dye. Solubilized QD sometimes interact with the dye, generating large signals external to cells that can be rinsed away. In this case the QDs give greater signals than any of the positive controls, despite lack of toxicity to cells. This occurs to different extents in different assays, perhaps reflecting the number of QDs that remain outside the cells when the dye is added or differences in the solubilization cap. It has a poor correlation with concentration, often showing maximal signal at very low QD concentrations (sub-nanomolar) Signal increases with irradiation, perhaps reflecting oxidation of the QD surface cap (Fig. 6.8).[23] For these reasons we do not recommend the use of this dye with QDs, especially those solubilized with a single functional molecule; it may be better with a thick polymer layer, although this remains to be verified.

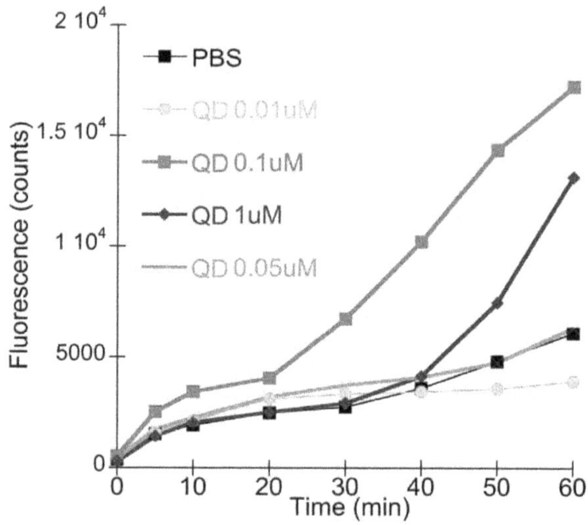

Figure 6.8 Dyes to watch out for #1: H_2DCFDA. Signal in plated cells with irradiation. The large signals seen with very low QD concentrations are erroneous and are many times that seen in positive controls. This false signal is not always seen.

Similarly, the commonly used live/dead stain combination (SYTO green/propidium iodide) can show alterations of both the green (live) and red (dead) channel when QDs are present, probably due to Förster resonance energy transfer (FRET) between the dyes and the QDs (Fig. 6.9). It is not possible to excite these dyes at wavelengths that do not excite the QDs. This is far from an exhaustive study of QD-dye interactions, and any reporter — especially a fluorescent one — should be tested for interactions with QDs before any firm conclusions are drawn.

6.6 Deliberate Toxicity: Photosensitized QDs

Since it was recognized early on that QDs could make ROS, efforts were made quite rapidly to exploit this as a feature for therapeutic applications.[24,25] Most anti-cancer drugs work by generating ROS within target cells, and ROS generation upon light exposure is the basis of *photodynamic therapy* (PDT). In PDT, a photosensitizer (or photocatalyst) is delivered to the region of interest. When it is sufficiently localized in the target zone, light (ultraviolet, visible, or infrared) makes the molecule able to transfer energy to another molecule; when the final acceptor is water or oxygen, this leads to the generation of ROS which is lethal to cells.[26] Certain cancers have been the primary targets of PDT. Porfimer sodium was approved for bladder cancer in 1993. Since then, hundreds of additional clinical trials have been launched. So far, all FDA-approved PDT photosensitizers absorb in the visible below 700 nm, limiting applications to lesions less than a few millimeters deep. This usually means skin cancer,[27] although any cancer that may be accessed by a fiber-optic cable and which is not excessively bulky may be treated this way; this includes lesions of the lung, breast, oesophagus, and even brain.[28]

There are compelling reasons to use PDT in malignant disease either alone or as an adjuvant. The treatment is cost-effective, highly localized and more specifically targeted to tumors than most chemotherapy agents, does not show the cumulative toxicity of radiation therapy and permits regeneration of normal tissue. PDT can even induce immunity which can contribute to elimination of tumor cells.[29,30] Nevertheless, PDT remains an alternative rather than a mainstay of cancer treatment. One of the greatest barriers to

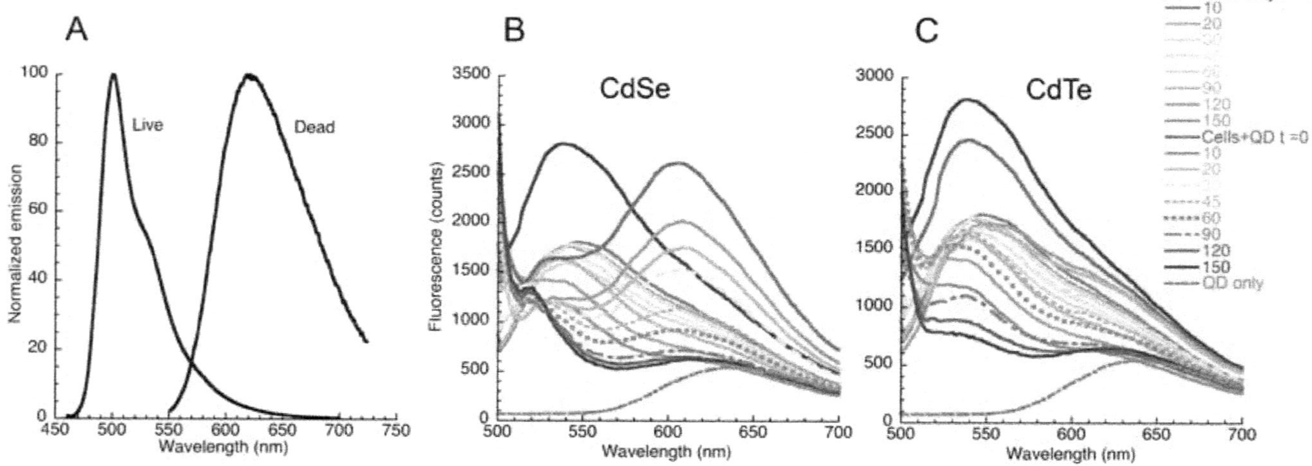

Figure 6.9 Dyes to watch out for #1: Live/Dead (SYTO9/propidium iodide). (A) Spectra of Live/Dead dyes alone, excited at 475 nm. (B) In the presence of red (peak ~640 nm) CdSe QDs. The cells alone show an intense Live signal that is slowly bleached upon irradiation. However, in the presence of the QDs, there is only an orange peak ~600 nm that does not represent either the QD or Dead emission. (C) CdTe QDs of the same color show different results. Although they seem more reasonable than those seen with CdSe, a true "Dead" peak is never seen.

its use is drug delivery. Photosensitizers are hydrophobic, and when they are delivered systemically, take several days to travel through the body to the tumor. During this time, the patient is sensitive to light and must stay out of the sun. Another serious issue is that for deeper tumors, including some skin cancers such as melanoma, the light does not penetrate the full depth of the tumor.

The use of PDT to destroy pathogenic bacteria and fungi is an area of research that has been recently re-awakened with the emergence of antibiotic resistance.[31,32] The idea goes back to Paul Ehrlich, who discovered that certain common bacterial stains such as methylene blue can act as photosensitizers. Although many fewer studies have been done using QDs for this sort of PDT than for anti-cancer PDT, the principles are the same, and any effective photosensitizing agent tested on mammalian cells might also be considered for anti-microbial PDT.

Nanoparticles in general have been called "an emerging paradigm" in PDT for a variety of reasons.[28] Nanoencapsulation of the drug molecules or conjugation of 10–100 molecules to hydrophilic nanoparticles has been shown in cell cultures and mice to provide a satisfactory solution to the problem of delivery. Several dozen studies have looked and different methods and materials for this purpose; one excellent example is gold. Gold nanoparticles 2–4 nm in diameter have been shown to aid in transport of PDT agents into cancer cells *in vitro*[33] and through the body in mice *in vivo*.[34] Although they do not generate ROS themselves, they aid in absorption of light, making the gold-bound drug approximately twice as toxic as a comparable level of free drug. While the conjugates have not yet received FDA approval, all of the ingredients have been approved for clinical applications (colloidal gold nanoparticles are approved for treatment of rheumatoid arthritis).

QDs have attracted a vast amount of interest as photodynamic therapy agents because they can act as photosensitizers — generate their own ROS — as well as act to improve the solubility and photophysics of traditional photosensitizers. Because of their broad absorbance spectra, their large extinction coefficients, and their large two-photon cross sections, they can be efficiently excited with the wavelength of choice, from the UV to the near-IR, and that they increase the overall light absorbance of a complex. QDs may interact with photosensitizer dyes by FRET or by direct electron transfer. In the latter case, the reactions are (S is the sensitizer):

$$S + e_{CB}^- \rightarrow S^{\bullet-}$$

$$O_2 + S^{\bullet-} \rightarrow O_2^{\bullet-} + S, \quad (6.6)$$

Or alternatively

$$S + h_{VB}^+ \rightarrow S^{\bullet+}$$

$$O_2 + S^{\bullet+} \rightarrow O_2^{\bullet-} + S\ (ox) \quad (6.7)$$

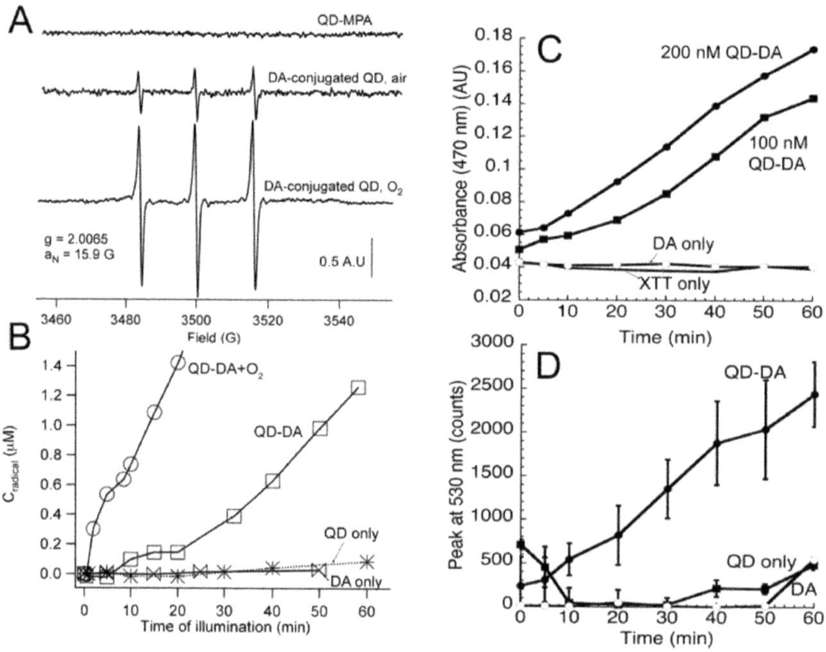

Figure 6.10 Singlet oxygen generation from QD-dopamine conjugates (QD-DA) as a function of blue light illumination. (A) Appearance of TMP radicals formed after 60 min of illumination. The QDs alone solubilized with MPA (QD-MPA) do not show any signal, whereas QD-DA shows a significant effect, which increases when oxygen is bubbled through the solution. (B) Concentration of TMP radical vs. irradiation time for QD-only, DA only, and QD-DA in air vs. bubbled with oxygen. (C) Signal with XTT, which are probably being reduced by superoxide (compare the kinetics with Fig. 6.7A). (D) Singlet oxygen sensor green shows a good deal of variability (error bars are SEM for three independent plates with triplicates in each). However, the only significant signal in 60 min is seen with QD-DA.

Increased ROS production compared with free drug has been shown with QDs conjugated to traditional PDT agents such as phthalocyanines,[25,35] merocyanine-540,[36] porphyrins,[37] rose bengal and chlorin,[38] as well as to other molecules such as platinum-based chemotherapeutic agents,[39] toluidine blue,[40] and dopamine.[7] Dopamine acts as in Eq. (6.7), and its presence on the QD leads to a disappearance of hydroxyl radical formation and an appearance of singlet oxygen in proportion to the amount of oxygen present in the solution (Fig. 6.10).

6.7 Toxicity to Bacteria: Introduction

Because of the well-known antimicrobial effect of certain nanomaterials, especially Ag, there are hundreds of studies examining the toxicity of nanoparticles to bacteria.[41] Bacterial studies have been performed with other types of semiconductor nanoparticles as well, especially TiO_2,[42–44] but also ZnO[45] and others including CuO.[46] Some comparisons of bacterial strains have been performed in these studies, usually comparing a Gram negative strain with a Gram positive. With all of these particles, mechanisms of toxicity are attributed to oxidation: either the photogeneration of ROS by the particles or the direct release of metal ions that can cause oxidative toxicity. Even with these materials, results of different studies vary greatly. In some studies the Gram positive test strain is reportedly more sensitive than the Gram negative;[43] in others it is the opposite.[47] These studies have been well reviewed.[48] Many mysteries still remain, and it is not clear whether the most important mechanisms have been identified in all cases.

Bacteria are different from eukaryotic cells in certain key ways that have a direct influence on their response to nanomaterials. First, while mammalian cells take up particles by endocytosis, bacteria do not endocytose. Possible mechanisms through which nanocrystals can pass through bacterial cell walls and membranes are nonspecific membrane damage, particle breakdown, and perhaps specific uptake through transporters or channels (although the latter has never been proven). Second, bacterial electron transport chains are located in their membranes, directly accessible to interactions with redox-active materials. Third, bacteria possess species- and strain-dependent mechanisms for dealing with heavy metals, including transporters to expel metal ions and mechanisms to "re-package" metals into nanoparticles of their own making. A schematic of possible QD interactions with a bacterial cell is given in Fig. 6.11.

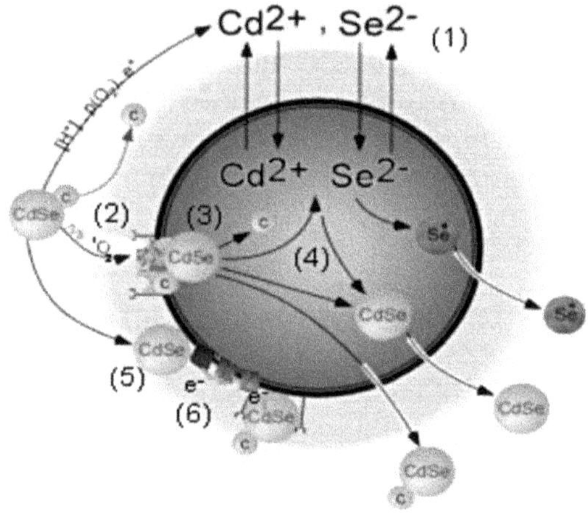

Figure 6.11 Possible mechanisms of QD interactions with bacteria. (1) Metal ions can pass through transporters and be specifically ejected from cells. These mechanisms are strain-dependent. Quantum dots outside the cell (2) can be broken down by light or weathering, causing the loss of their conjugates "c" and release of ions. They might also bind to cells (3) and may enter through damaged membranes. Once inside cells (4), QDs and metal ions may be further processed, repackaged, and sequestered or expelled. QDs that have lost their conjugates may interact with cells (5); QDs with or without conjugates might take electrons from the electron transport chain (6) or other membrane proteins. See also Colour Insert.

Perhaps most importantly, enzymes for detoxifying ROS can be very different in prokaryotic cells vs. eukaryotes. Many genes are involved, which can be up- and down-regulated depending upon environmental conditions; these mechanisms are only well-studied in a few strains, such as *E. coli*.[49] However, in such strains, generation of mutants lacking a single one of these genes is easily done, allowing for very specific tools to probe toxic mechanisms. The first key enzymatic reaction that protect cells from ROS is the dismutation of superoxide by the class of enzymes called **superoxide dismutases** (SOD):

$$M^{(n+1)+} - SOD + O_2^- \rightarrow M^{n+} - SOD + O_2$$

$$M^{n+} - SOD + O_2^- + 2H^+ \rightarrow M^{(n+1)+} - SOD + H_2O_2, \qquad (6.8)$$

where M is a metal ion (Cu^+, Mn^{2+}, Ni^{2+} or Fe^{2+}). Only the lactobacilli are known to have a different mechanism to protect against superoxide. The enzyme **catalase** catalyzes the decomposition of peroxide:

$$2H_2O_2 \rightarrow 2H_2O + O_2. \qquad (6.9)$$

Nearly all aerobic bacteria have catalase, as do a few anaerobes, but there are some notable exceptions, primarily the enterococci and the streptococci. Such bacteria do not produce bubbles when peroxide is added to them.

A third key enzyme in resistance to ROS is glucose 6-phosphate dehydrogenase (G6PD). G6PD catalyzes the first step in the pentose phosphate pathway, which provides reducing equivalents (NADPH) for reductive reactions and maintenance of the cellular redox state. NADPH is necessary for the function of enzymes that repair oxidative damage and regenerate antioxidant species.[50] Some *E. coli* NADPH reductases have been shown to reduce XTT,[51] so that the XTT assay may not be used as a measure of superoxide in bacteria, at least not this species and most likely others as well.

The few published studies on QDs are difficult to compare as they use different strains and very different types of QDs and conditions of exposure. However, there is agreement on certain points. Bacteria-related factors that are known to influence QD toxicity to bacteria include the following:

1. Gram sign;
2. Degree of QD binding to the bacteria, which is a function of Gram sign if the QDs are nonspecifically coated. This is unlikely to be electrostatic, as positively charged and negatively charged QDs bind the same strain in very similar quantities;
3. Genes for dealing with oxidative stress;
4. Heavy metal resistance genes.

One or more of these factors have been examined for each of the common types of QDs. We will discuss core CdSe, core/shell CdSe/ZnS (or CdS), and CdTe separately, as results for each are very different. Because of potential issues with dyes, we will focus on studies that evaluate toxicity by performing growth curves. This is usually done by measuring the optical density at 600 nm (OD600) to determine bacterial viability, followed by plate counts to confirm that the bacteria are able to form colonies. QDs added at quantities up to 100–200 nM will not absorb significantly at 600 nm regardless of their color; if very high concentrations of QDs are used, correction must be made for their absorbance.

6.8 CdSe Core QDs

A significant fraction of the studies on bacteria have been performed using CdSe bare core QDs. There are several reasons for this. One is that the weak fluorescence of CdSe core particles is less of a barrier to bacterial experiments than in those using larger cells, since most of the imaging is done using electron microscopy. In fact, some of the experiments have been done on CdSe QDs whose fluorescence is so weak it can't be observed! Second, CdSe QDs have dramatic, highly strain-dependent effects on bacteria that have attracted the interest of environmental toxicologists.

CdSe core QDs release significant Cd^{2+} ions when exposed to light or simply allowed to weather in aqueous solution for days to weeks.[21,52] This can be a source of toxicity to bacteria, including oxidative toxicity, resulting in DNA damage and lipid peroxidation. Cd^{2+} can also substitute for Zn^{2+} and cause interference with metabolic proteins.[53–55] Despite this, many bacterial strains are highly resistant to Cd^{2+}. In Gram positives, plasmids can encode for an expulsion system for Cd^{2+} ions, which are pumped out of the cell by an ATPase with a phospho-aspartate intermediate, which is encoded on a gene named *cadA*. In the case of Gram negative bacteria, Cd^{2+} (as well as Zn^{2+}, Ni^{2+}, and Co^{2+}) are expulsed by a three-polypeptide membrane complex that functions as a divalent cation/$2H^+$ antiporter.[56] In four common strains of bacteria, 2 Gram negative (*Pseudomonas aeruginosa*, *E. coli*) and 2 Gram positive (*Staphylococcus aureus*, *Bacillus subtilis*), we found EC50 concentrations of $CdCl_2$ to be in the low mM range for all except *Pseudomonas aeruginosa*, which was resistant to higher concentrations (Fig. 6.12). Se and Te compounds cause negligible toxicity to any of these strains up to the solubility limits of these compounds. In all of the strains, toxicity of CdSe QDs has been reported by us and others to be significantly greater than would be expected from the amount of Cd^{2+} released from the nanoparticles upon weathering.

There are several reasons for this. CdSe QDs cause significant membrane damage to bacterial cells, which can be observed by electron microscopy[57] (Fig. 6.13). Intact or nearly intact QDs

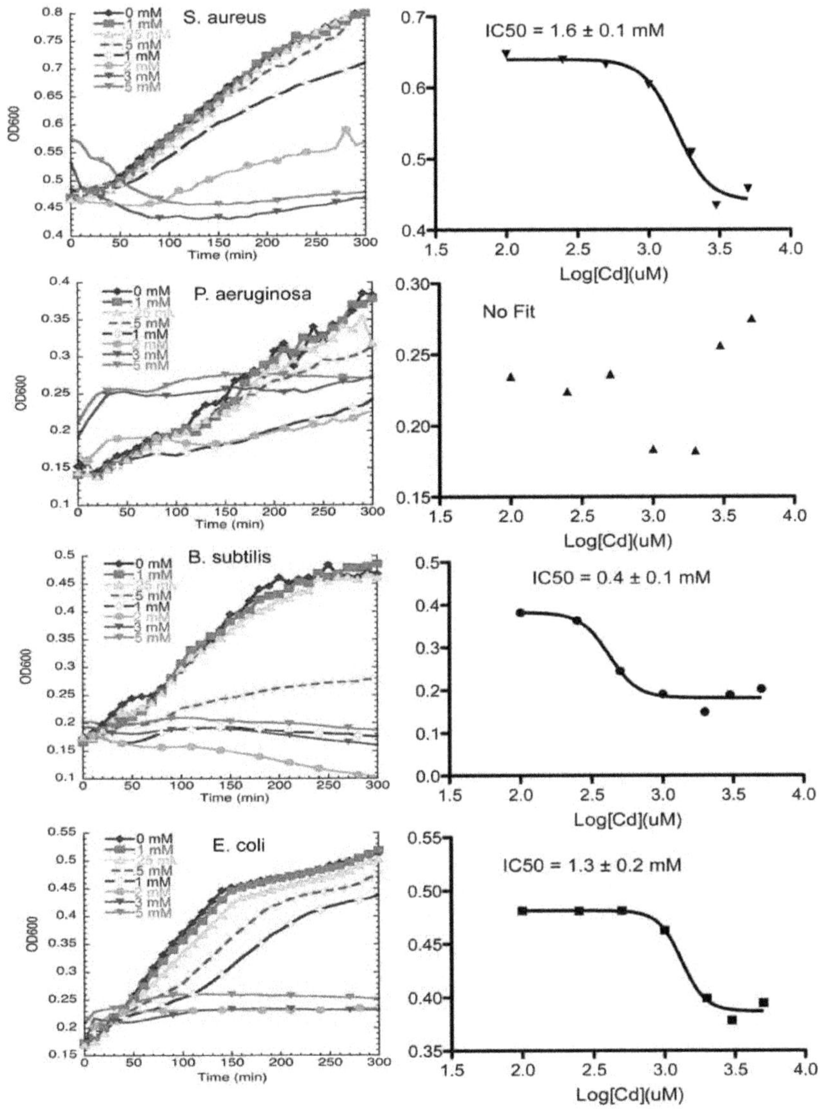

Figure 6.12 Growth curves of four strains with $CdCl_2$. Growth curves are a reliable way to measure toxicity to bacteria in the presence of fluorescent species such as QDs. The left panels show the OD600 values vs. time, recorded every 10 min, with different concentrations of the test substance. The right panels show the values at a chosen time point fit to the Hill equation of drug efficiency E vs. concentration C, $E = E_{max}\, C^x/(C^x + EC_{50}^x)$, in which E_{max} is the maximum obtainable effect and the EC_{50} the concentration at which E is 50% of E_{max}. The value x is the Hill coefficient.

(3–4 nm in core size or smaller) can then pass into the cells; if they are fluorescent enough to be imaged, they will show a broader fluorescence than the original QDs with evidence of collecting at the cell poles (Fig. 6.14). Once inside, the particles and their released ions can interact with intracellular structures, including DNA. CdSe QDs cause severe genotoxicity in *B. subtilis*, as can be observed from cellular elongation (Fig. 6.14A,B), observation of nucleoid destruction (Fig. 6.15), and measurements of oxidative DNA damage such as 8-hydroxydeoxy-guanosine ELISA assays.

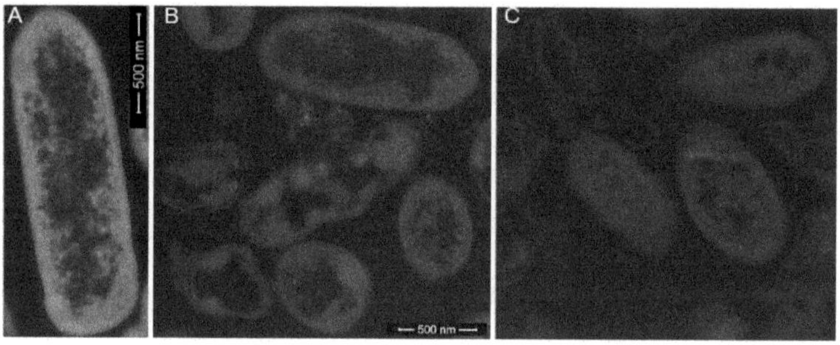

Figure 6.13 Scanning transmission electron microscopy (STEM) images of Pseudomonas aeruginosa. All specimens were stained with 2% Osmium tetroxide for 1 h. (A) Healthy cell incubated in rich medium. (B) Cells incubated with CdSe QDs to an equivalent concentration of 75 (micro)g/mL Cd. Note the broken, hollow "ghosts" of cells and destruction of internal cellular components. (C) Cells incubated with 75 (micro) g/mL cadmium acetate.

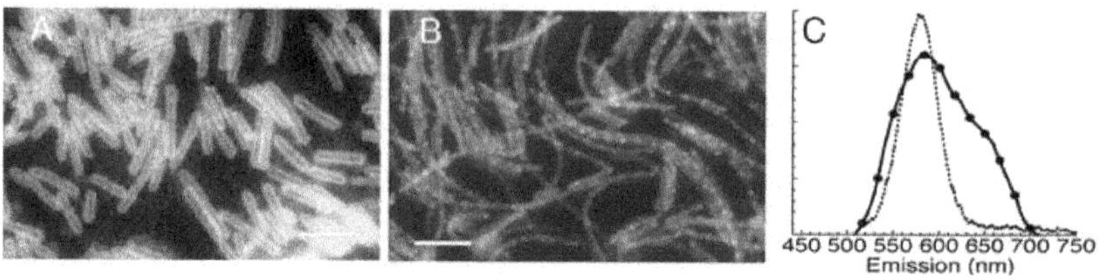

Figure 6.14 Quantum dot internalization by bacteria and observable toxic effects. (A) Control cells, wild-type *B. subtilis* labeled externally. Note the size and shape of the cells and lack of internal labeling. (B) CdSe QDs specifically targeted via adenine receptors and exposed to light. Note the broad range of colors, extra bright labeling near the poles of the cells, and marked cellular elongation. Elongation is a marker of genotoxicity, as the bacteria can grow but not reproduce. (C) Spectra of QDs before addition to bacteria (dashed), and after incubation with bacteria under light (solid). See also Colour Insert.

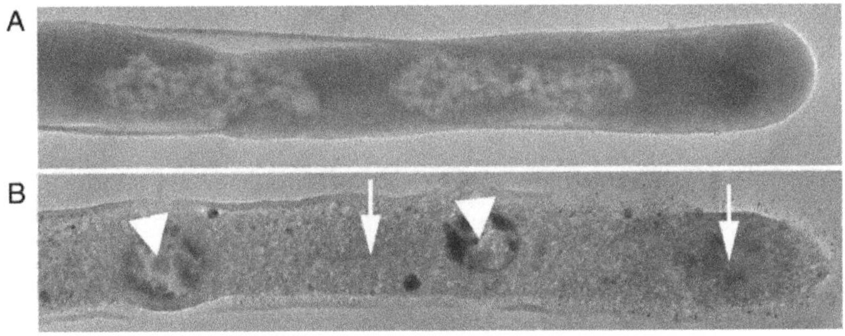

Figure 6.15 Observation of DNA destruction. TEM image of critical point dried *B. subtilis*. (A) Normal cells. Note signs of cell division and light-colored nucleoid region. (B) With targeted CdSe QDs exposed to light. The dark regions where the DNA should be are mostly carbon by x-ray spectroscopy (arrowheads); the Cd and Se are clustered at the poles (arrows) as in the fluorescence image.

All studies agree that both membrane damage and genotoxicity require tight binding to the cells.[52,57] This may be because the most reactive radicals, such as singlet oxygen and hydroxyl radicals, have short half-lives in solution so it is not surprising that they would need to be generated near cells to cause damage.The nanoparticles might also directly oxidize lipids and cell surface proteins, as has been suggested for fullerenes,[17] or cause other types of physical damage that are as yet unknown. Even highly metal-resistant strains show severe cell damage in response to QDs; one study found elevated intracellular ROS using H_2DCFDA in *P. aeruginosa*.[21]

Although detailed strain comparisons have not been done, there appears to be consensus that *B. subtilis* is more sensitive to CdSe than *E. coli*, and that *P. aeruginosa* is somewhat resistant (but all three of these strains have never been directly compared using the exact same QDs, so some skepticism is warranted!) This pattern follows the strains' relative sensitivity to heavy metals and so seems reasonable considering the breakdown of CdSe that occurs in solution. We have observed *E. coli* expelling large amounts of Cd-containing material after 1–2 h of incubation with CdSe QDs, followed by recovery of the growth curve (Fig. 6.16). This material was not in a 1:1 Cd:Se ratio and so did not represent intact QDs but rather breakdown products. In contrast, *B. subtilis* showed increasing toxicity after the 1–2 h time period.[57]

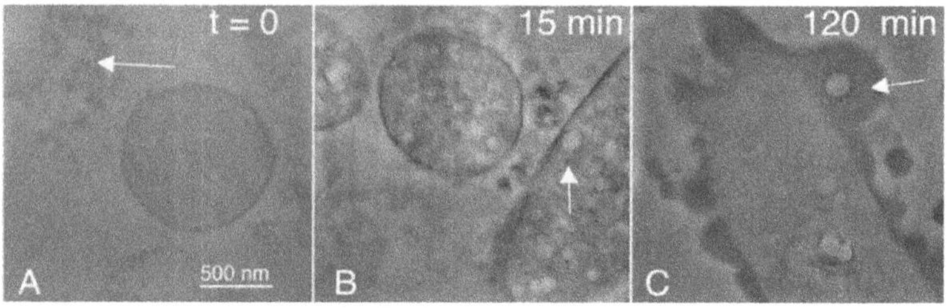

Figure 6.16 Cd uptake into and expulsion from *E. coli*. TEM images of thin sections of wild-type *E. coli* after exposure to targeted QD conjugates. For given amounts of time. Arrows indicate areas confirmed by X-ray spectroscopy to contain Cd and Se. (A) At time 0, the QDs are outside the cell (arrow) and the bacterium itself shows little contrast. (B) After 15 min of incubation under room light, the cell membrane is significantly darkened, and dark spots occur inside cells. This correlates with cell fluorescence. (C) After 2 h, most particles have disappeared from inside the cells, and dark areas containing Cd and Se are found outside the cell. This correlates with fluorescence loss and normal rates of cell reproduction.

Because the ROS-handling genes of *E. coli* are well known, there are ways to infer the types of ROS produced by examining the specific genes. One study used a panel of *E. coli* strains genetically engineered to express luciferase in response to a stressor of interest: peroxide (the *katG* gene), superoxides (the *micF* gene), DNA damage (the *recA* gene), and several others, exposed to CdSe QDs.[58] The authors found the greatest signal with the superoxide type of oxidative stress that was related to the length of the thiol chain used to solubilize the QDs (longer chain was more toxic). There was also a suggestion that small QDs were more toxic than larger QDs, although only two batches were tested. These experiments were done in the dark, so the toxicity could not have been due to photogenerated electrons or holes, but instead probably due to Cd leaching. The authors used energy filtered TEM (EFTEM) to map the elemental distributions around the treated cells, which showed that tight binding between the QDs and the cells occurred, probably with partial QD dissolution as reflected by a thick dark shell around many of the cells.

Although these studies are far from complete, they certainly show that CdSe QDs break down when incubated with bacteria in the presence or absence of light, but more quickly when irradiated. The breakdown products and the remaining QD core are both harmful to the cells and can cause membrane and DNA damage. The damage appears to be of the superoxide type in the dark, i.e. when it is Cd induced. The exact relationship between the length of the thiol cap and the QD core size and toxicity remain unknown, although smaller QDs appear to be more toxic under both light and dark conditions due possibly to tighter binding and/or the ability to pass into cells.

CdTe: Although CdTe QDs are usually used as bare core dots, thiol-capped CdTe QDs are much less reactive than thiol-coated CdSe and do not break down in solution nearly so quickly, and usually liberate sub-toxic levels of Cd^{2+} (nM or less) after 60 min or more of milliwatt-power light irradiation. They also remain highly fluorescent and useful for imaging for long periods of time. Several very thorough studies exploring the mechanisms of QD toxicity to one or more strains of bacteria have been published using these QDs. Despite the lack of Cd release, similar types of toxicity are seen with CdTe QDs as with CdSe, including cellular membrane defects and signs of oxidative stress. All studies agree that mechanisms are in some way oxidative, and most report EC50 values for *E. coli* of about 100 nM. The most remarkable difference from CdSe is that CdTe does not seem to be internalized by bacteria through membrane defects, at least not until the cells are completely destroyed.

One study used semi-quantitative rt-PCR to examine the expression levels of three genes involved in oxidative stress: *nfo* (encoding endonuclease IV, for DNA repair), *sod A* (encoding manganese-SOD), and *zwf* (encoding G6PD).[59] The authors found 50% downregulation of *nfo* and about 10% reduction for the other two genes. The same study also looked at SOD and peroxidase activity in treated cells, finding downregulation that was QD concentration-dependent. The authors also found tight binding between QDs and cells with surface morphological changes, but did not see any evidence of QD breakdown as measured by emission peak. Their studies were all performed without light exposure, although it was not specified whether the samples were always protected from room light. This seems to suggest that oxidative toxicity is possible with CdTe even in the dark. Only one color of QDs (orange) was used in these experiments.

In another study performed in our group, we used a variant of a recently developed assay using *E. coli* mutant strains, IC5204 (*oxy*R^-) and its control IC5282 (*oxy*R^+). *Oxy*R^- is sensitive to oxidative stress because of its deficiency in the *Oxy*R protein, a redox-sensitive transcriptional regulator which has a role in the activation of genes coding for antioxydant enzymes.[60] We see

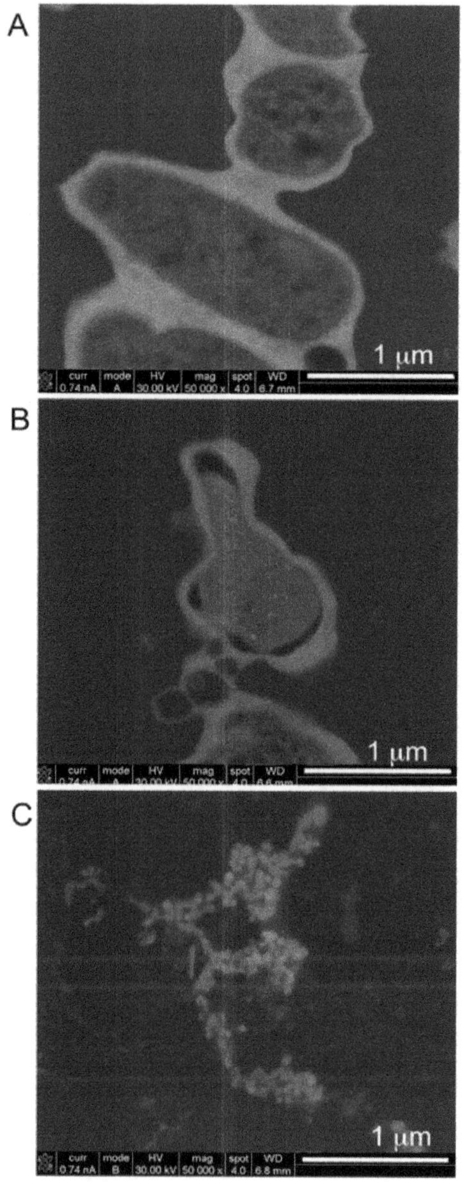

Figure 6.17 Destruction of *Oxy*R^{-}cells as observed with STEM. (A) Untreated cells. (B) Cells with CdTe and no light exposure. Note deformed cells with internalized particles. (C) After 30 min of 440 nm blue light, the cells are obliterated, leaving only fragments covered with nanoparticles.

significantly greater toxicity for CdTe in the *oxy*R⁻ strain by growth curve analysis, even without light exposure. With 30 min of 2 mW 440 nm blue light, 100 nM CdTe greatly inhibits the *oxy*R^- strain, but only slows the *oxy*R^+ strain. 500 nM CdTe kills the *oxy*R^- strain but only inhibits the *oxy*R^+ strain. It is important to note that light exposure alone has a moderate effect on the *oxy*R^- strain, reducing its growth rate significantly, but does not affect *oxy*R^+. This indicates that damaging species can be generated by light irradiation alone. Scanning transmission electron microscopy (STEM) illustrates the QD toxicity to the *oxy*R^- strain. Compared to untreated cells (Fig. 6.17A), CdTe-exposed cells show a distinct thinning of the exopolysaccharide (EPS) layer with internalization of nanoparticles and often deformation of the cell (Fig. 6.17B). After irradiation, the cells are completely destroyed (Fig. 6.17C).

Because of the remarkable finding that C60 toxicity in bacteria was due to direct oxidation of cell components, requiring tight binding to the cells, rather than ROS,[17] we looked for evidence of these processes with CdTe on four strains of bacteria.[61] The results are somewhat surprising. The strain most susceptible to the QDs, *E. coli*, is the one that shows the weakest binding; in fact there is an inverse correlation between degree of QDs bound to the cells after 4 washes and EC50 value (Fig. 6.18A). We see some evidence of membrane depolarization only in *S. aureus* and *B. subtilis*, to which the QDs bind tightly, and no evidence of depolarization in the more weakly bound Gram negative strains (Fig. 6.18B). Changes in the time constant of the QD fluorescence are also associated with tight QD binding, making this an interesting tool for examining QD-bacterial interactions (Fig. 6.18C). However, neither depolarization nor time constant change are associated with the greatest toxicity. The most sensitive strain is *E. coli*, which shows no membrane depolarization and no time constant change. The reactive species produced upon irradiation were oxidative species, as measured by EPR, and we believe these are the major source of toxicity under light, though the half-life of hydroxyl radicals is too short to affect bacteria that are not bound to QDs. It is unknown whether the unbound QDs contribute to the toxicity via other species, such as peroxide, or whether a small amount of bound QDs are sufficient to cause death to sensitive strains. Because depolarization is only weakly dependent upon light, this mechanism could help explain the non-light-related toxicity that every study has observed, although it is not applicable to *E. coli*.

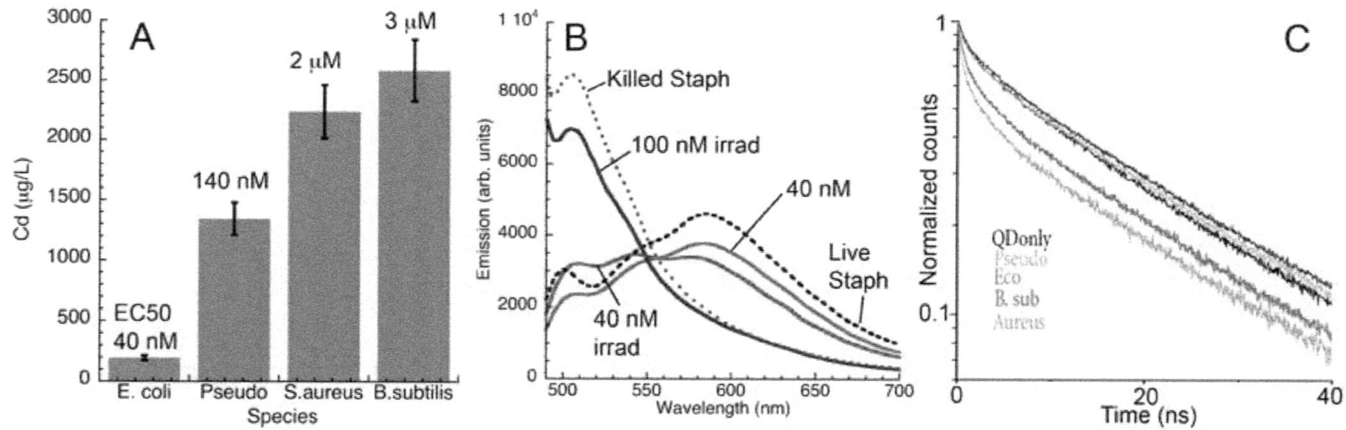

Figure 6.18 Binding of QDs to different bacterial strains and direct electron transfer. (A) Amount of Cd bound to different strains after four rounds of centrifugation and washing, as measured by atomic adsorption spectroscopy. There was no evidence of QD breakdown, so the amount of Cd should be directly proportional to the number of QDs. The value above each bar in the graph gives the EC50 for that strain. (B) Measurement of membrane potential using DiOC2(3). Live *Staphylococcus aureus* show a membrane potential as reflected by a red peak (585 nm). When killed by permeabilization or exposed to a depolarizing agent, the peak turns green (502 nm). The ratio between the two peaks gives a measure of membrane potential that is independent of bacterial concentration. (C) Time-resolved emission of QDs in the presence of different bacterial strains. Significant change in lifetime is seen only with the Gram positives.

Evidently, important questions remain. The relationship between CdTe size, thiol cap length, and toxicity has not been established. Our own results are contradictory; we see more variations between synthesis batches of the same size than between QDs of different sizes. EC50s can range from 10–100 nM in *E. coli* for different batches. The variations in toxicity in the dark are even greater. Many CdTe preparations show little inhibition in *E. coli* up to tens of nM in the dark; others show significant toxicity at low nM concentrations (Fig. 6.19). This implies some effect of the synthesis procedure that so far remains unknown. Different solubilization batches of the same synthesis batch are consistent with each other, so this is not function of cap exchange. The exact mechanisms of toxicity without light exposure need to be determined, as all of the ROS discussions in the previous sections imply photo-excitation. Until this is worked out, the exact mechanisms of CdTe antimicrobial activity remain a mystery.

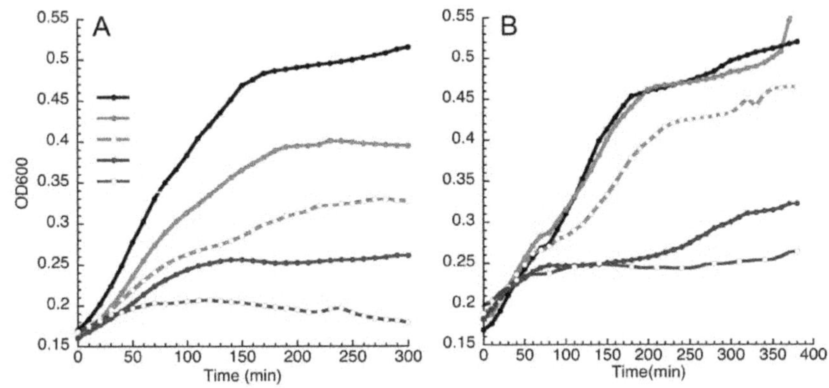

Figure 6.19 Differences in toxicity in the dark between two batches of orange-emitting CdTe QDs with *E. coli*. (A) Batch 1 shows substantial toxicity at very low QD concentrations in the dark. (B) Batch 2 shows no toxicity in the dark at the lowest concentrations, but substantial toxicity as the concentration rises.

6.9 CdSe/ZnS and Photosensitization

Only one study has reported significant toxicity of CdSe/ZnS nanoparticles to bacteria.[22] The authors used commercial CdSe/

ZnS coated with polyethylene glycol (PEG)-amine of PEG-carboxyl groups exposed to the metal-resistant species *Cupriavidus metallidurans*. Despite the different charges on these QDs, they associated with the bacteria to the same degree, suggesting that the binding was not electrostatic. Unfortunately, the article reports results based upon two dyes that are unreliable, both H_2DCFDA and propidium iodide. It is not known whether this study represents an artifact of the QD interaction with these dyes, or something particular to these commercial dots.

With our in-house prepared thiol-capped CdSe/ZnS QDs, we see negligible toxicity to our four common test strains. Light exposure in the presence of QDs causes no more cell death than irradiation alone (Fig. 6.20). Another study has confirmed lack of toxicity with CdSe/CdS core/shell QDs.[62] We therefore do not believe that thiol-solubilized core–shell QDs show significant bacterial toxicity.

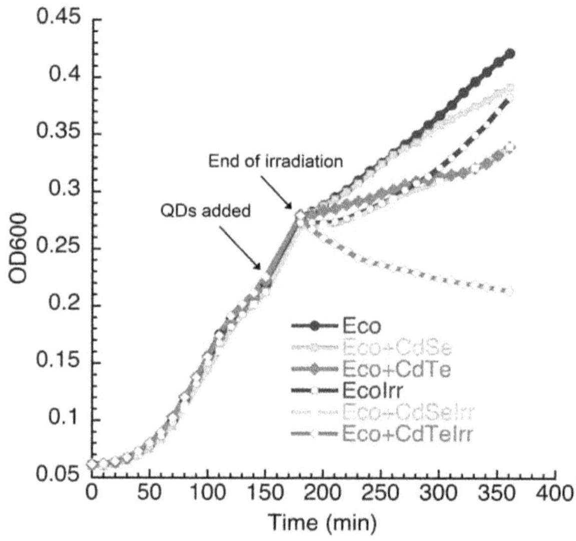

Figure 6.20 Effects of CdSe/ZnS vs. CdTe QDs on *E. coli* measured by growth curves. Substantial bacterial inhibition is only seen with CdTe and light exposure under these conditions (100 nM QDs, 30 min of 2.5 mW 440 nm light).

However, these particles may be photosensitized in order to create antimicrobial materials. One study used the commonly employed antimicrobial dye toluidine blue conjugated to CdSe/ZnS against streptococci and staphylococci.[63] The light used was 28 W white light from a fluorescent lamp. The authors found that at certain concentrations, QDs increased the anti-microbial effect of toluidine blue, although the QDs themselves caused no inhibition of bacterial growth. However, at the highest concentration of QDs, the mixture was less toxic than the dye alone. This emphasizes the importance of optimizing conjugation conditions for QD photosensitizers. What is also quite interesting is that this study reports singlet oxygen generation by direct measurement at 1270 nm. The authors found *less* singlet oxygen from the QD-toluidine blue mixture than from the dye alone, even when the mixture was more toxic to bacteria than the dye alone. The exact reasons for the increased toxicity of QD-toluidine blue is thus unknown, although it may be due to the production of other ROS species such as hydroxyl radicals.

We have also seen significant toxicity to bacteria using QD-dopamine only under light exposure, showing that the photosensitization observed in solution and with mammmalian cells is also applicable to bacterial toxicity. However, the binding of QD-dopamine to bacteria is very limited, so the effects are modest (Fig. 6.21).

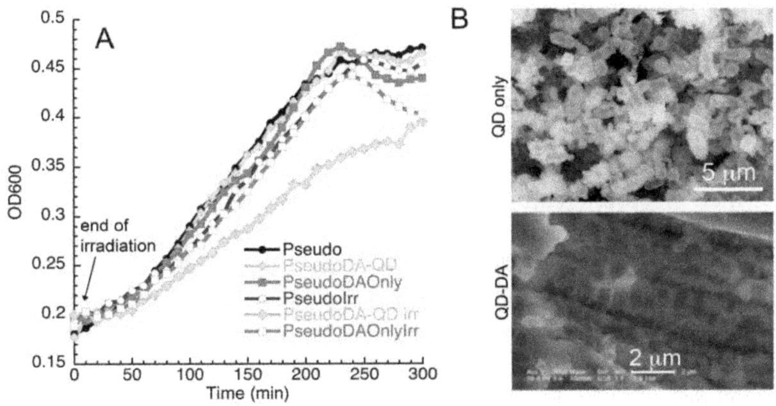

Figure 6.21 Effect of QD-dopamine on *Pseudomonas aeruginosa*. All QD concentrations are 100 nM. (A) Growth curves, starting after a period of QD addition ± 30 min blue light ("irr"), for bacteria alone, with dopamine alone, with QDs alone, and with QD-dopamine. The only significant effect is seen with QD-DA + irradiation. (B) QD-DA binds weakly to *P. aeruginosa*. These SEM images show QDs alone (top) and QD-DA (bottom) mixed with a bacterial culture, placed upon a stub, and rinsed once with H_2O.

6.10 Discussion

A very large number of factors are involved in QD toxicity to bacteria, including batch-to-batch synthesis variations that remain to be fully described. However, there is a consistent pattern of oxidative toxicity with all types of QDs, which may be Cd^{2+} mediated, occur without Cd^{2+} release but with photoexcitation of the QDs, or even occur in the dark. The mechanisms involved in all of these types of toxicity are different, making generalizations across different studies essentially impossible. A systematic study of light exposure and CdTe toxicity, for example, remains to be done. It also remains to elucidate the role of direct interactions, including direct electron transfer, in toxicty and potentially QD breakdown.

This is a field that deserves further study because the the final products of QD breakdown and repackaging are unknown. These may vary in the presence of different kinds of bacteria or other organisms, and they may have toxicity properties very different from those of the original QDs. The production of organic cadmium compounds or bacterially generated nanoparticles could pose a significant risk to the environment.[64] The final goal behind many of the experiments reviewed here is to develop rational recommendations for the treatment and release of nanoparticles into the soil or water, which can only be done by addressing the possible interactions with target organisms.

As the experimental parameters for testing become better defined in liquid culture, it will be important to move to more complex, realistic systems. Many bacteria exist in biofilms, and their interactions with nanoparticles will be altered by differences in surface area, permeability, and gene expression relative to planktonic cultures. The way differently coated QDs permeate through these types of media remains an important question.[48]

There is also the possibility that QDs could be developed as antimicrobials, either for *in vitro* use (e.g. on surfaces, in wastewater treatment plants) or for *in vivo* applications (e.g. topical creams for infected wounds). For these applications, it is critical to maximize the antimicrobial properties of QDs while eliminating the possibility of human or animal exposure to Cd^{2+}. Many studies support the idea of QDs as photosensitizers for photodynamic therapy, both anti-cancer and anti-microbial. There is also a vast body of literature on the use of TiO_2 nanoparticles for photo-disinfection. QDs have the advantage over TiO_2 of absorbing visible light rather than UV; in fact, TiO_2 itself may be sensitized by QDs. The promise of the particles seems vast, but two primary questions remain. First, can QDs be passivated sufficiently to prevent Cd^{2+} release without destroying their ROS-generating processes? And second, can Cd-free materials with the same optical properties as QDs serve as photosensitizers? The synthesis of InP particles with emission spectra spanning the visible range has been reported. This material has a bandgap similar to that of CdTe. However, InP is very sensitive to water and oxygen destruction, and thick ZnS shells are usually needed to prevent rapid breakdown. Under these conditions, photosensitization is difficult because energy and electron transfer are impaired.

This chapter aimed to introduce only the most common materials and surface chemistries as a means to illustrate the tools that exist for investigating the physics, chemistry, and biology of nanoparticle/cell interactions. The possible combinations of materials, functionalization, and biological system are nearly limitless. The challenge for the future is to identify the guiding general principles that will allow us to design nanomaterials for a huge variety of applications while minimizing risks to human health and the environment. Bacteria can serve as a biological test system, but also show many unique features that can be exploited for materials synthesis, decontamination, and biosensing. The range of possibilities has only begun to be explored.

References

1. Maurer-Jones, M.A., Bantz, K.C., Love, S.A., Marquis, B.J., and Haynes, C.L. Toxicity of therapeutic nanoparticles. *Nanomed*, **4**, 219–241 (2009).
2. Marquis, B.J., Love, S.A., Braun, K.L., and Haynes, C.L. Analytical methods to assess nanoparticle toxicity. *Analyst*, **134**, 425–439 (2009).
3. Jan, E., Byrne, S.J., Cuddihy, M., Davies, A.M., Volkov, Y., Gun'ko, Y.K., and Kotov, N.A. High-content screening as a universal tool for fingerprinting of cytotoxicity of nanoparticles. *ACS Nano*, **2**, 928–938 (2008).
4. Kovochich, M., Espinasse, B., Auffan, M., Hotze, E.M., Wessel, L., Xia, T., Nel, A.E., and Wiesner, M.R. Comparative toxicity of C60 aggregates toward mammalian cells: role of tetrahydrofuran (THF) decomposition. *Environ Sci Technol*, **43**, 6378–6384 (2009).
5. Hoshino, A., Fujioka, K., Oku, T., Suga, M., Sasaki, Y.F., Ohta, T., Yasuhara, M., Suzuki, K., and Yamamoto, K. Physicochemical properties and

cellular toxicity of nanocrystal quantum dots depend on their surface modification. *Nano Lett*, **4**, 2163–2169 (2004).

6. Derfus, A.M., Chan, W.C.W., and Bhatia, S.N. Probing the cytotoxicity of semiconductor quantum dots. *Nano Lett*, **4**, 11–18 (2004).
7. Clarke, S.J., Hollmann, C.A., Zhang, Z., Suffern, D., Bradforth, S.E., Dimitrijevic, N.M., Minarik, W.G., and Nadeau, J.L. Photophysics of dopamine-modified quantum dots and effects on biological systems. *Nat Mater*, **5**, 409–417 (2006).
8. Medintz, I.L., Uyeda, H.T., Goldman, E.R., and Mattoussi, H. Quantum dot bioconjugates for imaging, labelling and sensing. *Nat Mater*, **4**, 435–446 (2005).
9. Clift, M.J., Rothen-Rutishauser, B., Brown, D.M., Duffin, R., Donaldson, K., Proudfoot, L., Guy, K., and Stone, V. The impact of different nanoparticle surface chemistry and size on uptake and toxicity in a murine macrophage cell line. *Toxicol Appl Pharmacol*, **232**, 418–427 (2008).
10. Schneider, R., Wolpert, C., Guilloteau, H., Balan, L., Lambert, J., and Merlin, C. The exposure of bacteria to CdTe-core quantum dots: the importance of surface chemistry on cytotoxicity. *Nanotechnology*, **20**, 225101 (2009).
11. Hardman, R. A toxicologic review of quantum dots: toxicity depends on physicochemical and environmental factors. *Environ Health Perspect*, **114**, 165–172 (2006).
12. Ryman-Rasmussen, J.P., Riviere, J.E., and Monteiro-Riviere, N.A. Surface coatings determine cytotoxicity and irritation potential of quantum dot nanoparticles in epidermal keratinocytes. *J Invest Dermatol*, **127**, 143–153 (2007).
13. Wuister, S.F., Donega, C., and Meijerink, A. Influence of Thiol Capping on the Exciton Luminescence and Decay Kinetics of CdTe and CdSe Quantum Dots. *J Phys Chem B*, **108**, 17393–17397 (2004).
14. Burda, C., Green, T.C., Link, S., and El-Sayed, M.A. Electron shuttling across the interface of CdSe nanoparticles monitored by femtosecond laser spectroscopy. *J Phys Chem B*, **103**, 1783–1788 (1999).
15. Ipe, B.I., Lehnig, M., and Niemeyer, C.M. On the generation of free radical species from quantum dots. *Small*, **1**, 706–709 (2005).
16. Cho, S.J., Maysinger, D., Jain, M., Roder, B., Hackbarth, S., and Winnik, F.M. Long-term exposure to CdTe quantum dots causes functional impairments in live cells. *Langmuir*, **23**, 1974–1980 (2007).

17. Lyon, D.Y., Brunet, L., Hinkal, G.W., Wiesner, M.R., and Alvarez, P.J. Antibacterial activity of fullerene water suspensions (nC60) is not due to ROS-mediated damage. *Nano Lett*, **8**, 1539–1543 (2008).

18. Lovric, J., Cho, S.J., Winnik, F.M., and Maysinger, D. Unmodified cadmium telluride quantum dots induce reactive oxygen species formation leading to multiple organelle damage and cell death. *Chem Biol*, **12**, 1227–1234 (2005).

19. Chen, J.Y., Lee, Y.M., Zhao, D., Mak, N.K., Wong, R.N., Chan, W.H., and Cheung, N.H. Quantum dot-mediated photoproduction of reactive oxygen species for cancer cell annihilation. *Photochem Photobiol* (2009).

20. Kim, J., Park, Y., Yoon, T.H., Yoon, C.S., and Choi, K. Phototoxicity of CdSe/ZnSe quantum dots with surface coatings of 3-mercaptopropionic acid or tri-*n*-octylphosphine oxide/gum arabic in Daphnia magna under environmentally relevant UV-B light. *Aquat Toxicol*, **97**, 116–124 (2010).

21. Priester, J.H., Stoimenov, P.K., Mielke, R.E., Webb, S.M., Ehrhardt, C., Zhang, J.P., Stucky, G.D., and Holden, P.A. Effects of soluble cadmium salts versus CdSe quantum dots on the growth of planktonic *Pseudomonas aeruginosa*. *Environ Sci Technol*, **43**, 2589–2594 (2009).

22. Slaveykova, V.I., Startchev, K., and Roberts, J. Amine- and Carboxyl-Quantum Dots Affect Membrane Integrity of Bacterium Cupriavidus metallidurans CH34. *Environ Sci Technol*, **43**, 5117–5122 (2009).

23. Cooper, D.R., Dimitrijevic, N.M., and Nadeau, J.L. Photosensitization of CdSe/ZnS QDs and reliability of assays for reactive oxygen species production. *Nanoscale*, **2**, 114–121 (2010).

24. Bakalova, R., Ohba, H., Zhelev, Z., Ishikawa, M., and Baba, Y. Quantum dots as photosensitizers? *Nat Biotechnol*, **22**, 1360–1361 (2004).

25. Samia, A.C.S., Chen, X.B., and Burda, C. Semiconductor quantum dots for photodynamic therapy. *J Am Chem Soc*, **125**, 15736–15737 (2003).

26. Calzavara-Pinton, P.G., Venturini, M., and Sala, R. Photodynamic therapy: update 2006 — Part 1: Photochemistry and photobiology. *J Eur Acad Dermatol Venereol*, **21**, 293–302 (2007).

27. Devirgiliis, V., Panasiti, V., Curzio, M., Gobbi, S., Rossi, M., Roberti, V., and Calvieri, S. Complete remission of nodular basal cell carcinoma after combined treatment with photodynamic therapy and imiquimod 5% cream. *Dermatol Online J*, **14**, 25 (2008).

28. Chatterjee, D.K., Fong, L.S., and Zhang, Y. Nanoparticles in photodynamic therapy: an emerging paradigm. *Adv Drug Deliv Rev*, **60**, 1627–1637 (2008).

29. Thong, P.S., Olivo, M., Kho, K.W., Bhuvaneswari, R., Chin, W.W., Ong, K.W., and Soo, K.C. Immune response against angiosarcoma following lower fluence rate clinical photodynamic therapy. *J Environ Pathol Toxicol Oncol*, **27**, 35–42 (2008).

30. Preise, D., Oren, R., Glinert, I., Kalchenko, V., Jung, S., Scherz, A., and Salomon, Y. Systemic antitumor protection by vascular-targeted photodynamic therapy involves cellular and humoral immunity. *Cancer Immunol Immunother*, **58**, 71–84 (2009).

31. Perussi, J.R. Photodynamic inactivation of microorganisms. *Quimica Nova*, **30**, 988–994 (2007).

32. Wainwright, M. Photodynamic antimicrobial chemotherapy (PACT). *J Antimicrob Chemother*, **42**, 13–28 (1998).

33. Oo, M.K., Yang, X., Du, H., and Wang, H. 5-aminolevulinic acid-conjugated gold nanoparticles for photodynamic therapy of cancer. *Nanomed*, **3**, 777–786 (2008).

34. Cheng, Y., A, C.S., Meyers, J.D., Panagopoulos, I., Fei, B., and Burda, C. Highly efficient drug delivery with gold nanoparticle vectors for in vivo photodynamic therapy of cancer. *J Am Chem Soc*, **130**, 10643–10647 (2008).

35. Ma, J., Chen, J.Y., Idowu, M., and Nyokong, T. Generation of singlet oxygen via the composites of water-soluble thiol-capped CdTe quantum dots – Sulfonated aluminum phthalocyanines. *J Phys Chem B*, **112**, 4465–4469 (2008).

36. Narayanan, S., Sinha, S., and Pal, S. Sensitized emission from chemotherapeutic drug conjugated to CdSe/ZnS QDs. *J Phys Chem C*, **112**, 12716–12720 (2008).

37. Shi, L.X., Hernandez, B., and Selke, M. Singlet oxygen generation from water-soluble quantum dot-organic dye nanocomposites. *J Am Chem Soc*, **128**, 6278–6279 (2006).

38. Tsay, J.M., Trzoss, M., Shi, L., Kong, X., Selke, M., Jung, M.E., and Weiss, S. Singlet oxygen production by Peptide-coated quantum dot-photosensitizer conjugates. *J Am Chem Soc*, **129**, 6865–6871 (2007).

39. Blanco, N.G., Maldonado, C.R., and Mareque-Rivas, J.C. Effective photoreduction of a Pt(IV) complex with quantum dots: a feasible new light-induced method of releasing anticancer Pt(II) drugs. *Chem Commun* (*Camb*), 5257–5259 (2009).

40. Narband, N., Mubarak, M., Ready, D., Parkin, I., Nair, S., Green, M., Beeby, A., and Wilson, M. Quantum dots as enhancers of the efficacy of bacterial lethal photosensitization. *Nanotechnology*, **19**, 445102 (2008).

41. Morones, J.R., Elechiguerra, J.L., Camacho, A., Holt, K., Kouri, J.B., Ramirez, J.T., and Yacaman, M.J. The bactericidal effect of silver nanoparticles. *Nanotechnology*, **16**, 2346–2353 (2005).

42. Mitoraj, D., Janczyk, A., Strus, M., Kisch, H., Stochel, G., Heczko, P.B., and Macyk, W. Visible light inactivation of bacteria and fungi by modified titanium dioxide. *Photochem Photobiol Sci*, **6**, 642–648 (2007).

43. Fu, G., Vary, P.S., and Lin, C.T. Anatase TiO_2 nanocomposites for antimicrobial coatings. *J Phys Chem B*, **109**, 8889–8898 (2005).

44. Chung, C.J., Lin, H.I., Tsou, H.K., Shi, Z.Y., and He, J.L. An antimicrobial TiO_2 coating for reducing hospital-acquired infection. *J Biomed Mater Res B Appl Biomater*, **85**, 220–224 (2008).

45. Nair, S., Sasidharan, A., Divya Rani, V.V., Menon, D., Manzoor, K., and Raina, S. Role of size scale of ZnO nanoparticles and microparticles on toxicity toward bacteria and osteoblast cancer cells. *J Mater Sci Mater Med*, **20 Suppl 1**, S235–241 (2009).

46. Heinlaan, M., Ivask, A., Blinova, I., Dubourguier, H.C., and Kahru, A. Toxicity of nanosized and bulk ZnO, CuO and TiO_2 to bacteria Vibrio fischeri and crustaceans Daphnia magna and Thamnocephalus platyurus. *Chemosphere*, **71**, 1308–1316 (2008).

47. Rincon, A.G., and Pulgarin, C. Use of coaxial photocatalytic reactor (CAPHORE) in the TiO_2 photo-assisted treatment of mixed Escherichia coli and Bacillus subtilis and the bacterial community present in wastewater. *Catal. Today*, **101**, 331–344 (2005).

48. Neal, A.L. What can be inferred from bacterium-nanoparticle interactions about the potential consequences of environmental exposure to nanoparticles? *Ecotoxicology*, **17**, 362–371 (2008).

49. Lushchak, V.I. Oxidative stress and mechanisms of protection against it in bacteria. *Biochemistry (Mosc)*, **66**, 476–489 (2001).

50. Lundberg, B.E., Wolf, R.E., Jr., Dinauer, M.C., Xu, Y., and Fang, F.C. Glucose 6-phosphate dehydrogenase is required for *Salmonella typhimurium* virulence and resistance to reactive oxygen and nitrogen intermediates. *Infect Immun*, **67**, 436–438 (1999).

51. Benov, L., and Fridovich, I. Is reduction of the sulfonated tetrazolium 2,3-bis (2-methoxy-4-nitro-5-sulfophenyl)-2-tetrazolium 5-carboxanilide a reliable measure of intracellular superoxide production? *Anal Biochem*, **310**, 186–190 (2002).

52. Mahendra, S., Zhu, H.G., Colvin, V.L., and Alvarez, P.J. Quantum dot weathering results in microbial toxicity. *Environ Sci Technol*, **42**, 9424–9430 (2008).

53. Nies, D.H. Microbial heavy-metal resistance. *Appl Microbiol Biotechnol*, **51**, 730–750 (1999).

54. Hartwig, A., Asmuss, M., Ehleben, I., Herzer, U., Kostelac, D., Pelzer, A., Schwerdtle, T., and Burkle, A. Interference by toxic metal ions with DNA repair processes and cell cycle control: molecular mechanisms. *Environ Health Perspect*, **110 Suppl 5**, 797–799 (2002).

55. Smeets, K., Cuypers, A., Lambrechts, A., Semane, B., Hoet, P., Van Laere, A., and Vangronsveld, J. Induction of oxidative stress and antioxidative mechanisms in *Phaseolus vulgaris* after Cd application. *Plant Physiol Biochem*, **43**, 437–444 (2005).

56. Silver, S. Bacterial resistances to toxic metal ions — A review. *Gene*, **179**, 9–19 (1996).

57. Kloepfer, J.A., Mielke, R.E., and Nadeau, J.L. Uptake of CdSe and CdSe/ZnS quantum dots into bacteria via purine-dependent mechanisms. *Appl Environ Microbiol*, **71**, 2548–2557 (2005).

58. Senevirathna, W., Kiro, R., Rosen, R., Popov, I., Belkin, S., and Wells, M. CdSe quantum dots induce superoxide stress in engineered biosensor bacteria. *Nanotoxicology*, **3**, 98–108 (2009).

59. Lu, Z.S., Li, C.M., Bao, H.F., Qiao, Y., Toh, Y.H., and Yang, X. Mechanism of antimicrobial activity of CdTe quantum dots. *Langmuir*, **24**, 5445–5452 (2008).

60. Martinez, A., Urios, A., and Blanco, M. Mutagenicity of 80 chemicals in Escherichia coli tester strains IC203, deficient in OxyR, and its oxyR(+) parent WP2 uvrA/pKM101: detection of 31 oxidative mutagens. *Mutat Res*, **467**, 41–53 (2000).

61. Dumas, E., Gao, C., Suffern, D., Bradforth, S.E., Dimitrijevic, N.M., and Nadeau, J.L. Interfacial charge transfer between CdTe quantum dots and Gram negative vs Gram positive bacteria. *Environ Sci Technol*, **44**, 1464–1470 (2010).

62. Aruguete, D.M., Guest, J.S., Yu, W.W., Love, N.G., and Hochella, M.F. Interaction of CdSe/CdS core-shell quantum dots and *Pseudomonas aeruginosa*. *Environ Chem*, **7**, 28–35 (2009).

63. Narband, N., Mubarak, M., Ready, D., Parkin, I.P., Nair, S.P., Green, M.A., Beeby, A., and Wilson, M. Quantum dots as enhancers of the efficacy of bacterial lethal photosensitization. *Nanotechnology*, **19** (2008).

64. Aruguete, D.M., and Hochella, M.F. Bacteria-nanoparticle interactions and their environmental implications. *Environ Chem*, **7**, 3–9 (2009).

65. Pryor, W.A. Oxy-radicals and related species: their formation, lifetimes, and reactions. *Annu Rev Physiol*, **48**, 657–667 (1986).

66. Wood, P.M. The potential diagram for oxygen at pH 7. *Biochem. J.*, **253**, 287–289 (1988).

67. Rajendran, V., Lehnig, M., and Niemeyer, C.M. Photocatalytic activity of colloidal CdS nanoparticles with different capping ligands. *J Mater Chem*, **19**, 6348–6353 (2009).

68. Dumas, E., Gao, C., Suffern, D., Bradforth, S.E., Dimitrijevic, N.M., and Nadeau, J.L. Interfacial charge transfer between CdTe quantum dots and Gram negative vs Gram positive bacteria. *Environ Sci Technol*, **44**, 1464–1470 (2010).

69. Hagfeldt, A., and Gratzel, M. Light-induced redox reactions in nanocrystalline systems. *Chem. Rev.*, **95**, 49–68 (1995).

Index

Colour Insert

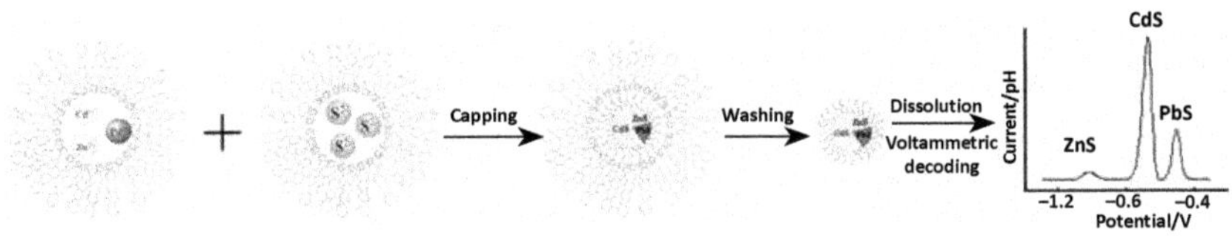

Figure 1.9

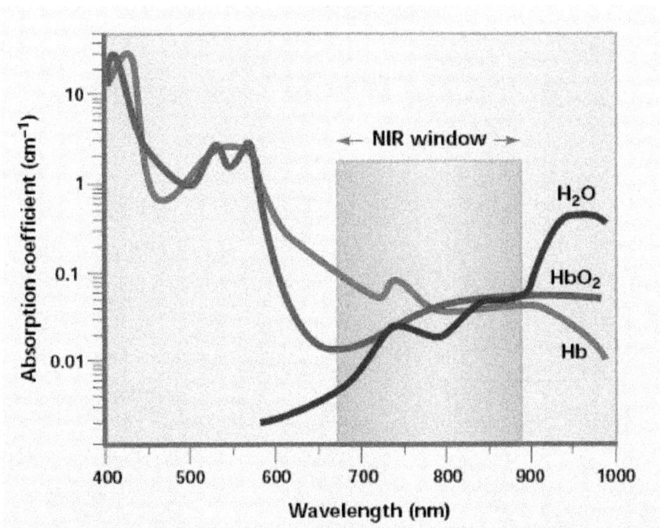

Figure 1.15

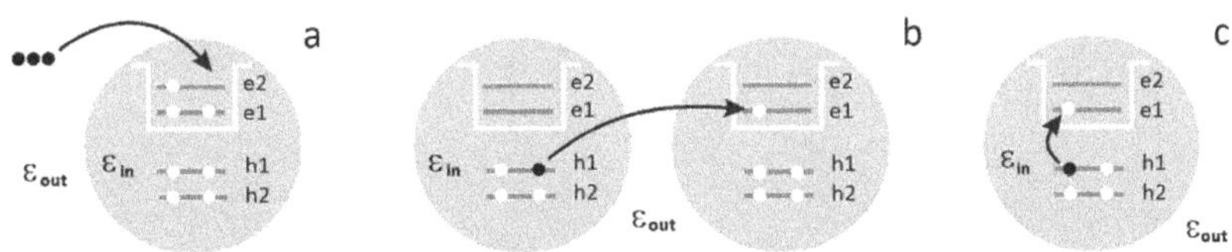

Figure 3.1

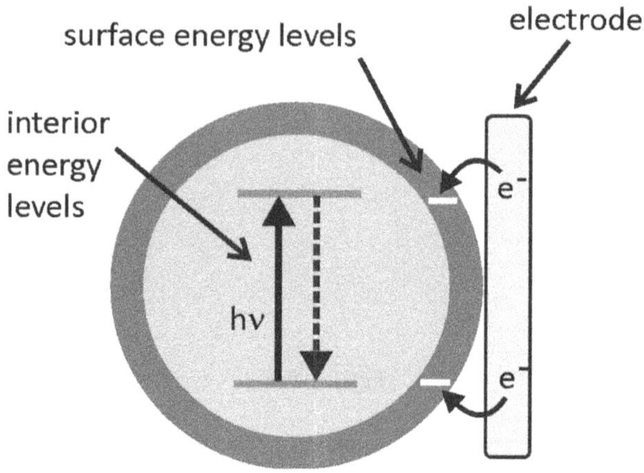

Figure 3.2

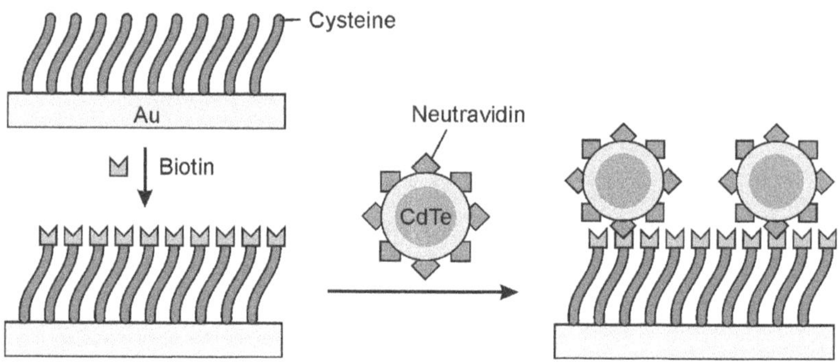

Figure 3.17

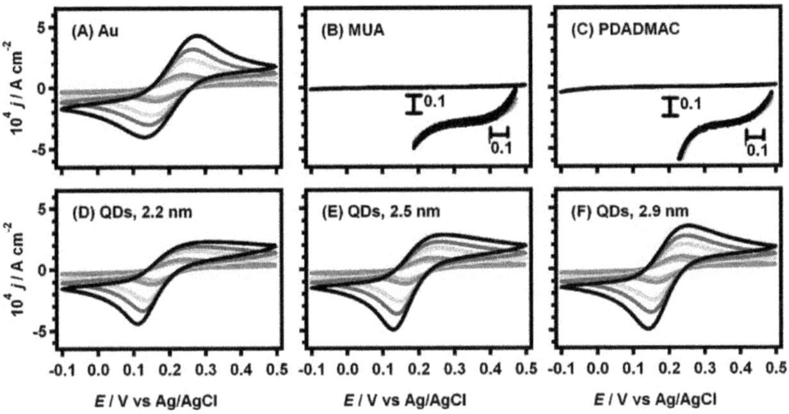

Figure 3.19

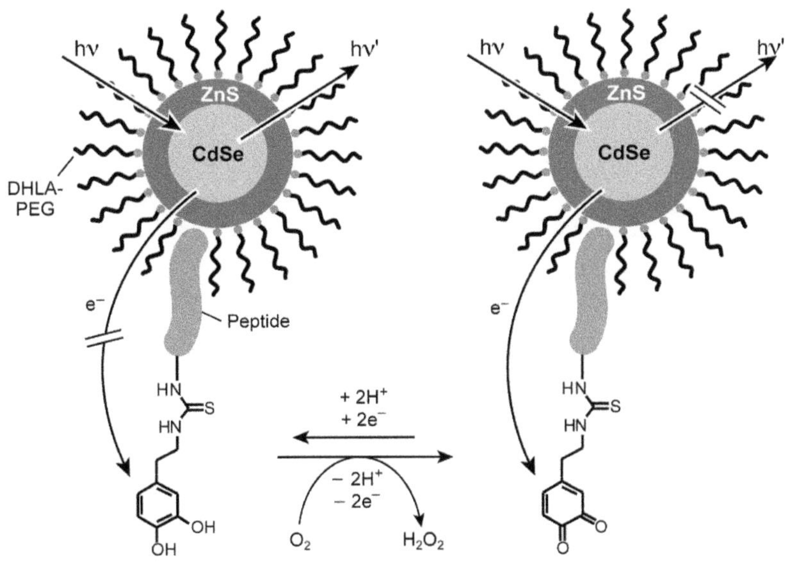

Figure 3.20

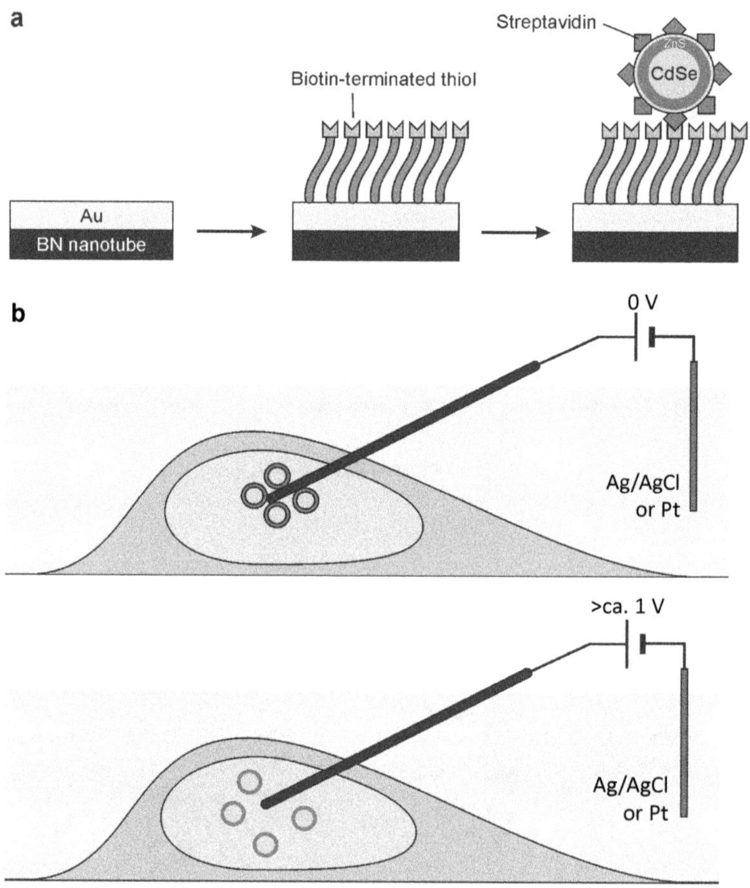

Figure 3.21

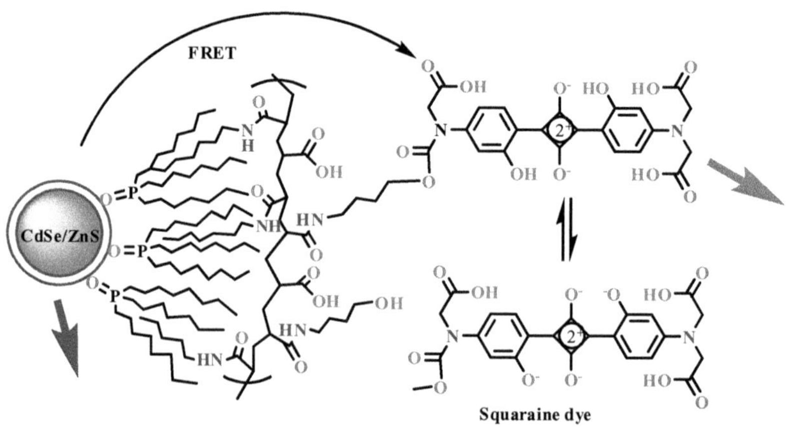

Figure 5.2

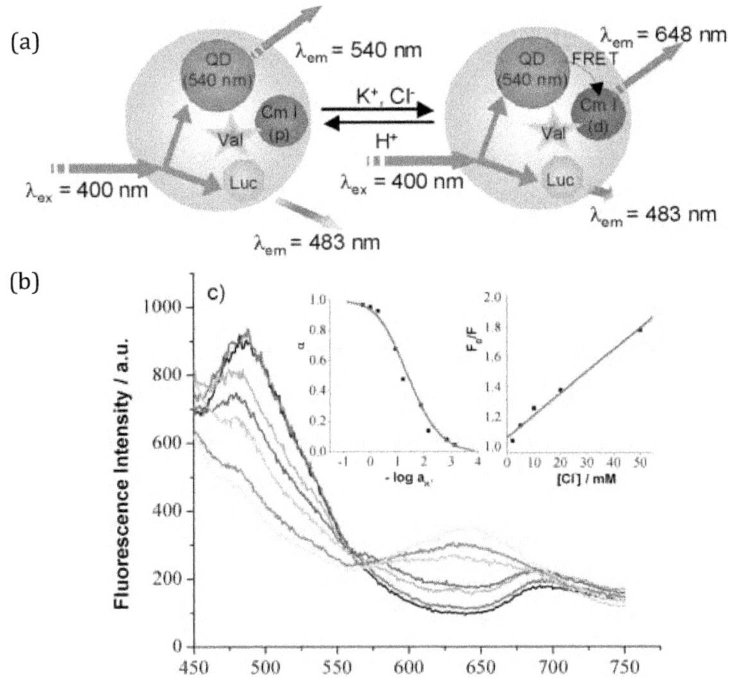

Figure 5.5

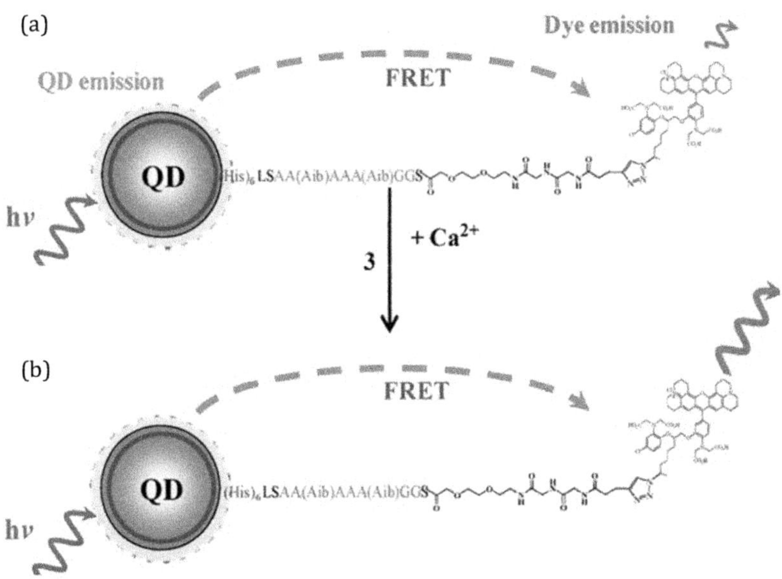

Figure 5.6

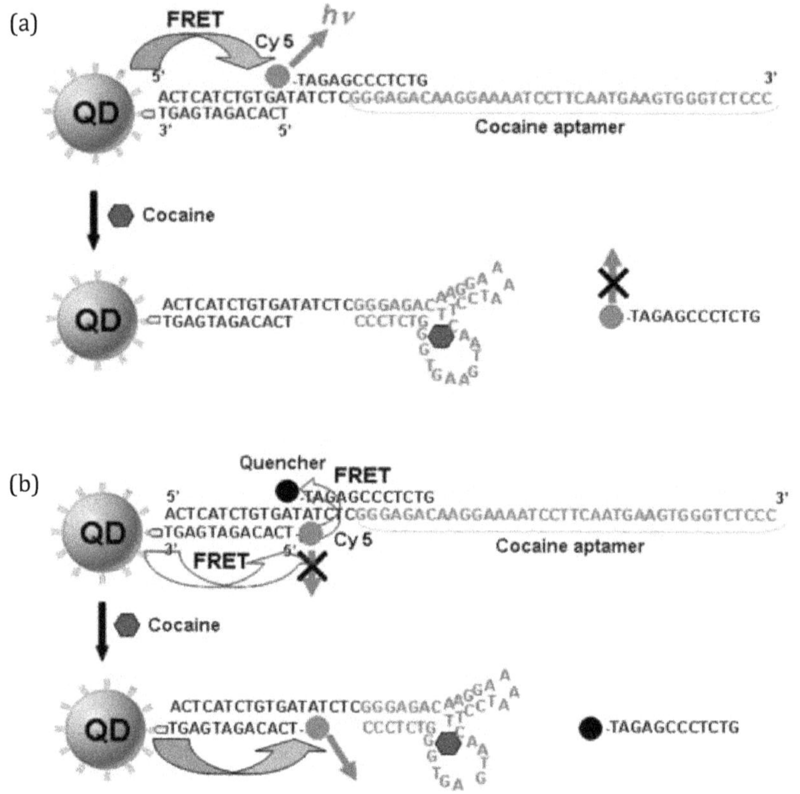

Figure 5.7

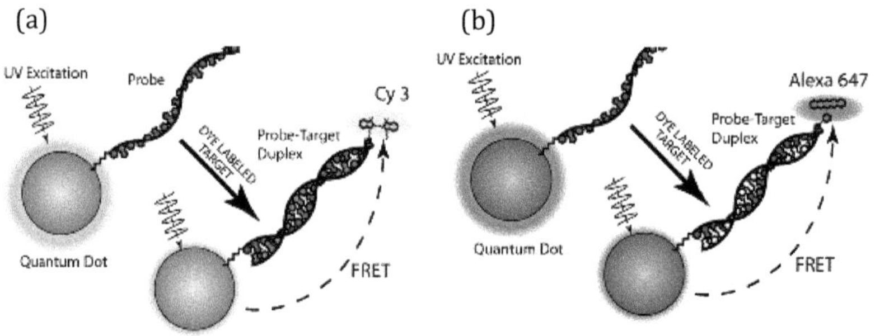

Figure 5.12

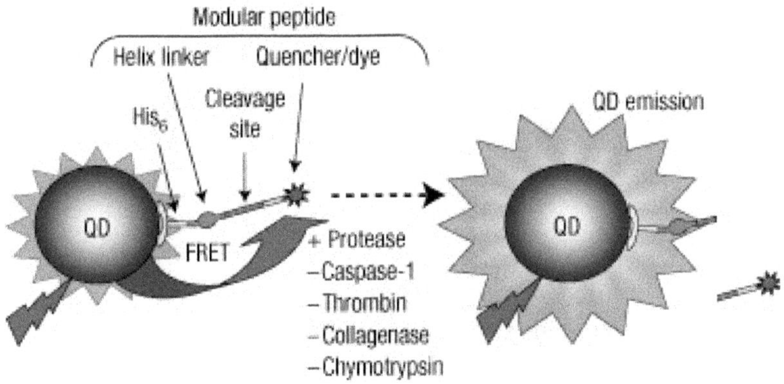

Figure 5.16

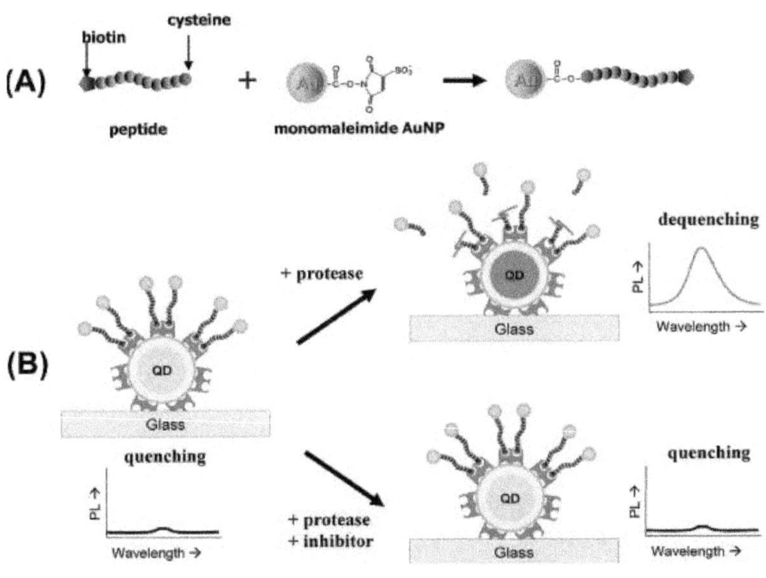

Figure 5.18

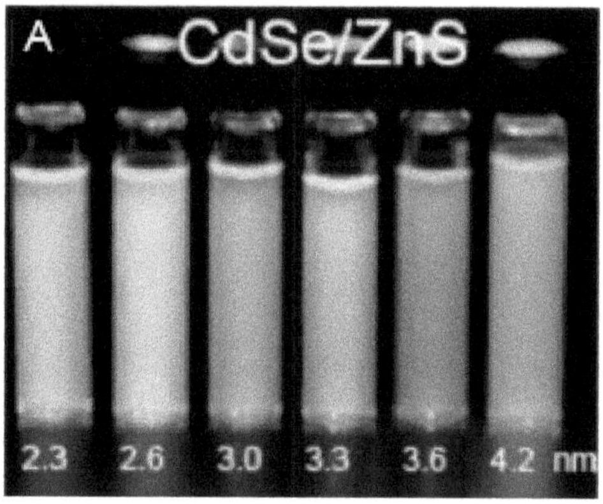

Figure 6.1

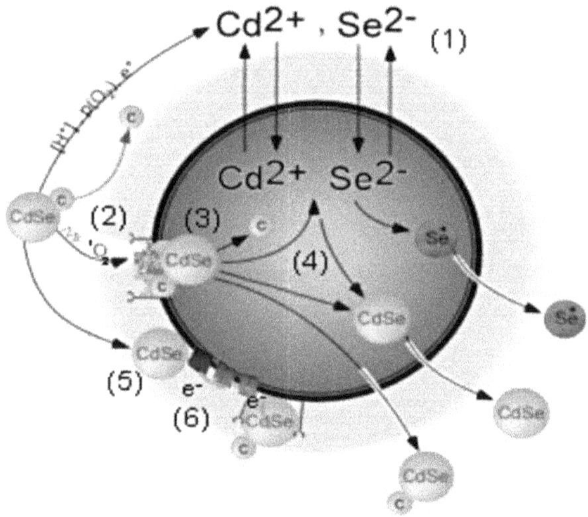

Figure 6.11

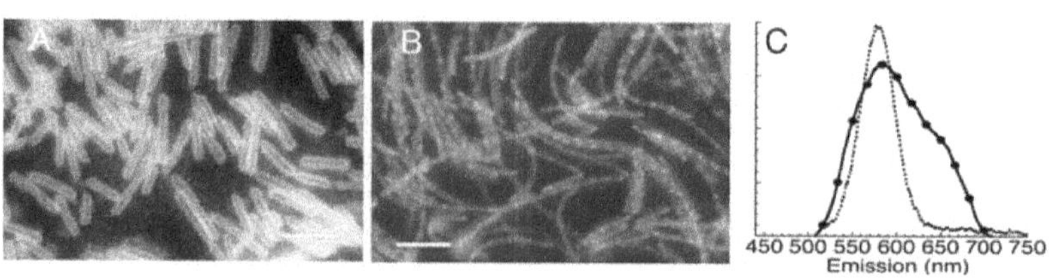

Figure 6.14

Milton Keynes UK
Ingram Content Group UK Ltd.
UKHW020748221123
433026UK00004B/8

9 789814 316002